SPECIAL FORCES

• • • • • • • •

SPECIAL FORCES

BRUCE QUARRIE

CHARTWELL
BOOKS, INC.

A QUINTET BOOK

Published by Chartwell Books
A Division of Book Sales, Inc.
110 Enterprise Avenue
Secaucus, New Jersey 07094

ISBN 1–55521–575–0

This book was designed and produced by
Quintet Publishing Limited
6 Blundell Street
London N7 9BH

Creative Director: Peter Bridgewater
Art Director: Ian Hunt
Designer: James Lawrence

Typeset in Great Britain by
Central Southern Typesetters, Eastbourne
Manufactured in Singapore by
Tien Wah Press (Pte) Ltd

Contents

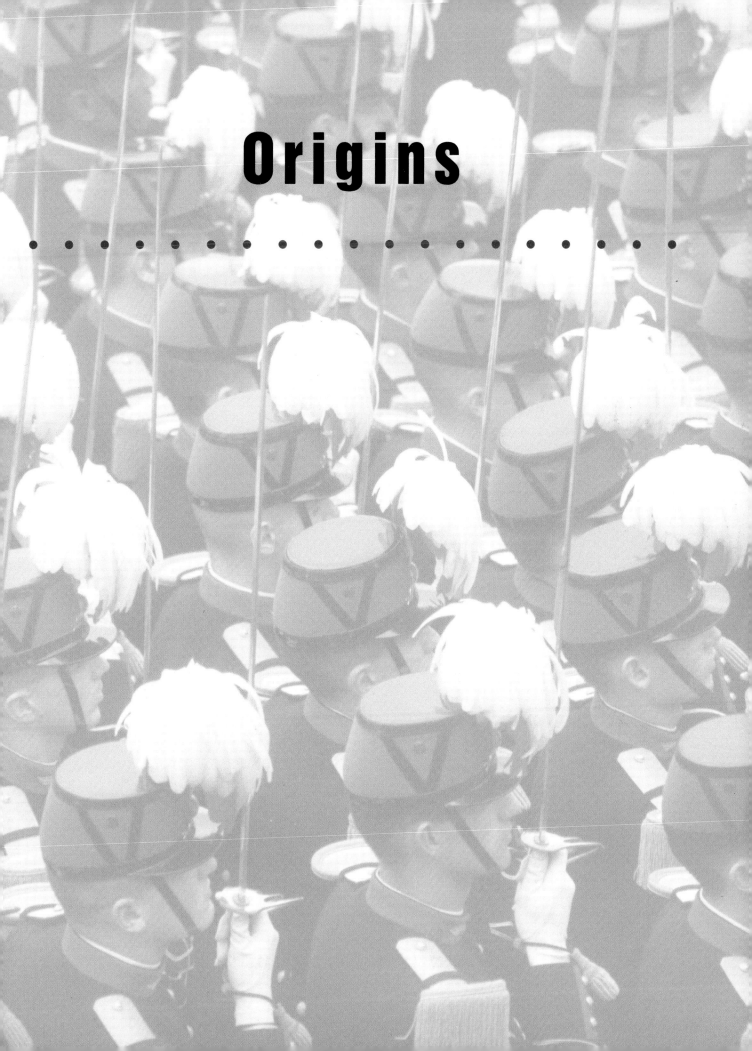

Origins

• • • • • • • • • • • • • • • • • • •

The tradition of elite military forces stretches in an unbroken line from the Praetorian Guard of the Caesars through Napoleon's Imperial Old Guard to Britain's Brigade of Guards and other modern equivalents. Within every army in history there have been one or two units which, because of superior fighting skill, loyalty and discipline, have been considered a 'cut above' the rest. On the battlefield such troops would generally be held in reserve ready either to administer the coup de grâce when the enemy was on the point of defeat, or to act as an intact rallying point for the remainder of the army in case things went wrong. During and since the Second World War, however, a different type of elite force has emerged which recruits from within the existing elite, and it is with these that this book is concerned.

During the First World War, when the fighting on the Western Front bogged down in the stalemate of the trenches, the German army pioneered the use of small raiding parties. These were compact groups of veteran volunteers who, unencumbered with all the infantryman's heavy kit but armed to the teeth instead with close-quarter weapons, would sneak through the barbed wire at night to fall on an unsuspecting stretch of Allied trench, kill everyone in sight and retreat back into no man's land before reserves could be rushed to the spot. It was for the use of such troops that the first sub-machine gun was developed.

These Stosstruppen, or stormtroopers, as they were called, formed the model for the 'kommandos' of Hitler's feared Waffen-SS. Under Otto Skorzeny, their exploits included the daring rescue of Mussolini when he was being held by Italian partisans in a mountaintop hotel, and a lightning raid on Tito's headquarters in Yugoslavia. This only failed because the leader of the communist resistance was not there at the time. They also kidnapped Hungarian Regent Admiral Horthy's son from a fortress in the heart of Budapest to stiffen the Hungarian government's resolve to continue fighting the Russians. Skorzeny's 150th Brigade also took part in the Battle of the Bulge at the end of 1944 when many of its men, dressed in American army uniforms and driving American vehicles, were caught and subsequently shot as spies. Ironically, one of the two principal SS officer training schools in what is now West Germany, Bad Tölz, is today used by the American Rangers!

The German army and air force created their own special forces apart from these Waffen-SS commandos. Italy had pioneered the use of the parachute as a method of dropping troops into battle in the 1920s and the Soviet Union had eagerly adopted the idea. German observers were particularly impressed by a Soviet exercise near Kiev in 1935 in which 2,500 paratroopers dropped to secure an airfield and hold it while reinforcements were flown in. As a direct result, the Germans spurred on their researches into military gliders and formed their first parachute battalion in January 1936. By 1940 there were five battalions, which played a major role in the conquest of Norway in April and France and the Low Countries in May, dropping ahead of the regular ground forces to secure bridges and airfields. They also landed on top of the supposedly impregnable Belgian fortress of Eben Emael, whose garrison was so stunned that a thousand men surrendered with hardly a fight to a mere 85 men, only six German soldiers being killed. German paras subsequently went on to fight in North Africa, Italy and Russia, their most famous achievement being the aerial conquest of Crete in May 1941.

The German army had also realized the importance of special forces to race ahead and seize key objectives in advance of the Panzer divisions in their planned Blitzkrieg ('lightning war'). At the beginning of 1939 the head of the Abwehr (military intelligence), Admiral Wilhelm Canaris, formed a company of predominantly Polish-speaking volunteers who, dressed in Polish uniforms, seized key bridges and rail junctions in Poland on 1 September that year. This tiny unit eventually grew to the size of a regiment, the 'Brandenburg' Regiment, and carried out many similar operations during the invasions of Belgium in 1940 and Russia in 1941. Modelling themselves on Lawrence of Arabia, who was one of the earliest modern exponents of unconventional warfare, in 1942 a small party of Brandenburgers made an epic 2,000-mile trek across the Sahara to escort a pair of German agents into Cairo, and in 1943 the regiment spearheaded the assault on the island of Leros in the Aegean.

Unfortunately for them, the necessities of war eventually saw the cream of the German elite forces frittered away in ordinary infantry actions as the Allies chipped away at the frontiers of the Third Reich.

The Second World War also saw the raising of similar units in the armies of other countries. Immediately after the fall of France in June 1940 a Colonel on the British General Staff, Dudley Clarke, pressed for a reorganization of the earlier Independent Companies which had been created in the spring of the 'Phoney War' for behind-the-lines sabotage operations. No 3 Commando, led by Lieutenant-Colonel John Durnford-Slater, was the result, the name 'commando' deriving from the groups of irregular soldiers who had fought the Boers in South Africa at the turn of the century. What would have been No 1 Commando was formed from the Independent Companies and, five battalions strong, was originally given the rather unfortunate name Special Service (SS) Brigade, although this was abandoned during 1941. Similarly, the original No 2 Commando formed the nucleus of the new British airborne force which Prime Minister Winston Churchill had ordered created on 22 June 1940.

During the course of the war the number of army commando battalions was expanded to 12 (there was also a 13th, superstitiously numbered No 14 Commando, for a brief time), plus a further three in the Middle East (Nos 50, 51 and 52 [ME]). The Royal Marines also formed nine between 1942 and 1944, numbered 40 to 48 (RM), and more than 18,000 army and Marine Commandos took part in the invasion of Normandy in 1944. Nos 10 and 51 Commandos were the most unusual. The first was an interallied force of mixed nationalities which even included a number of anti-Nazi German volunteers, serving under false identities in case of capture. The latter was a mixed force of Palestinian Arab and Jewish volunteers with British officers which principally fought in East Africa.

One obvious criterion for any elite force is that its members should all be volunteers because the nature of their tasks demands a very high level of motivation and self-discipline. It was in recognition of these qualities that the wearing of the distinctive green Commando beret was authorized after the Dieppe raid in August 1942.

Meanwhile, what would have been No 2 Commando had mustered at Ringway, near Manchester, in July 1940 and in November was renamed No 11 Special Air Service Battalion. This title was dropped in September 1941 and the unit became 1st Battalion, 1st Parachute Brigade, commanded by Lieutenant-Colonel E E 'Dracula' Down. The 2nd Battalion was originally commanded by Lieutenant-Colonel E W C Flavell and subsequently by an officer whose name has become synonymous with wartime parachute exploits, Major John Frost. The 3rd Battalion was led by Lieutenant-Colonel G W Lathbury.

Then came the formation of the Glider Pilot Regiment in December as a further step towards the creation of the 1st Airborne Division which Churchill had demanded after the loss of Crete had demonstrated what parachute and glider forces could achieve. In August 1942 the para battalions became The Parachute

Regiment and, envious of the Commandos' green beret, they adopted a maroon one as their own trademark. By the time of D-Day there were two complete divisions, the 1st commanded by Major-General Roy Urquhart and the 6th under Major-General Richard Gale. Overall commander of airborne forces was Brigadier F A M 'Boy' Browning.

The commandos and the airborne forces established formidable reputations during the Second World War and their modern successors have continued the tradition, as we shall see. There is no space here to describe their wartime exploits – the Lofoten Islands' raid, Vaagso, St Nazaire (which resulted in the award of five Victoria Crosses), Dieppe, the invasions of French North-West Africa and Sicily, Normandy, Arnhem, Comacchio, etc, not to mention smaller-scale adventures such as Bruneval, when a combined airborne/commando force raided a German radar installation in northern France to bring its secrets back for the 'boffins'. There are many fine books on these operations, several written by the men who actually took part in them.

Going back to 1941, however, we find the 'Special Air Service' (SAS) title revived in North Africa – this time permanently. The early history of what is widely regarded as the ultimate elite military formation of modern times deserves special attention. Nos 7, 8 and 11 Commandos had been sent to Egypt in the spring of that year to help counter the German forces under Erwin Rommel, despatched by Hitler to aid his hapless ally, Mussolini. They were generally known as 'Layforce' after their commander, Major-General Robert Laycock, but were soon split up, No 11 going to Cyprus while No 7 joined 50 and 52 (ME) Commandos on Crete. Most of the men of the latter three units were killed or captured fighting a valiant rearguard action when the German paras landed in May, which just left No 8 in Egypt. One of its junior officers was Lieutenant David Stirling.

Kicking his heels wondering what to do after the fall of Crete, he and three friends decided to try their own hands at parachuting. Unfortunately, Stirling's parachute failed to deploy properly on his first jump and he had to spend several weeks in hospital. While languishing there he evolved the concept of tiny raiding parties of no more than four or five men, which would be far more elusive than a regular Commando company of 50 or so. These could be dropped by parachute or infiltrated by Jeeps or small boats, deep behind enemy lines to carry out reconnaissance or sabotage missions. When he left hospital Stirling managed to persuade General Sir Claude Auchinleck, C-in-C in the Western Desert, of the validly of his ideas, which remain the cornerstone of all special

PREVIOUS PAGE Colonel (now Sir) David Stirling (right) with a section of SAS men in their famous 'Pink Panther' Jeeps in North Africa. ABOVE A corporal of the Long Range Desert Group nurses the twin Lewis machine-guns in his Jeep. Chevrolet trucks were also widely used.

forces to this day. The result was 'L' Detachment, Special Air Service Brigade. There was, in fact, no longer an SAS Battalion, let alone Brigade. The name was a ploy to make Rommel think that a strong British parachute force had arrived in North Africa. The name stuck and soon there were two full SAS regiments, the 1st under Major R B 'Paddy' Mayne (after David Stirling was captured in January 1943), the 2nd by Stirling's brother Bill.

Stirling's SAS shared many exploits with the Long Range Desert Group (formed in December 1940 by Major Ralph Bagnold) and 'Popski's Private Army', the force of Libyan Arab commandos led by Belgian-born expatriate Russian Major Vladimir Peniakoff. Their officers and men freely exchanged ideas and evolved hard-hitting tactics for behind-the-lines operations against German and Italian airfields and fuel dumps. These tactics would be honed and refined over the years in many battles around the world, particularly in the modern fight against terrorism in which the SAS is the acknowledged master. It was in the camaraderie of this environment that the enlisted men of the SAS learned to call their officers by the now cherished title 'boss', which soon puts any Sandhurst superman in his place! The present day Royal Marines' Special Boat Squadron is also part of the early story for it originally evolved from a tiny component of 'Layforce' as part of Stirling's inaugural SAS. The British created dozens of similar small units for longer or shorter periods during the war but only the SAS and SBS have retained their identities and the two – elites within an elite – maintain close links alongside the usual friendly rivalry.

After fighting in the desert and later carrying out many daring penetration missions in Italy, the two SAS regiments were returned to the UK and brigaded together with a Belgian squadron and two Free French parachute battalions to form the 1st Special Air Service Brigade – now a reality rather than a codename. Working closely with SOE – the Strategic Operations Executive – and resistance groups in occupied Europe, they carried out dozens of invaluable reconnaissance and sabotage missions both before and after D-Day, their teams often operating 50 miles and more ahead of the front line in constant dread of detection.

Entering the war later than the UK, the United States was, surprisingly, quicker to begin to develop similar forces. The first plans for a test platoon of paratroopers were approved in April 1940 but it was not until July, following the fall of France, that things really started moving. Even then progress was slow because America still hoped to stay out of another European war. The platoon did not make its first drop from an aircraft until August and establishment of a proper training school at Fort Benning did not occur until the following April. By the time of the Japanese attack on Pearl Harbor there was still only one operational battalion, the 501st. Their jump word of 'Geronimo!' remains part of the US airborne forces' tradition to this day.

The spring of 1942 saw a huge leap forward and the creation of another 26 consecutively numbered battalions, many of whose officers and men visited England to witness British training methods and results and to meet their allied opposite numbers. By August two embryo divisions existed, the battalions having grown to regimental strength. These were the 82nd ('All Americans') commanded initially by Major-General Omar N Bradley and subsequently by Brigadier-General Matthew B Ridgway, and the 101st ('Screaming Eagles') commanded by Lieutenant-Colonel (later Brigadier-General) William C Lee. Their first operation came when a battalion of 503rd Regiment of the 82nd parachuted in to capture an airfield in Tunisia following Operation 'Torch', the Anglo-American invasion of French North-West Africa in November 1942. Elements of both divisions took part in many subsequent operations in Sicily and Italy before being withdrawn to England for D-Day when, alongside the British 6th Airborne Division, they landed on the French coast in advance of the main amphibious assault.

They subsequently took part in the bold attempt to cross the Rhine codenamed 'Market Garden', the 82nd landing to secure bridges at Nijmegen and the 101st at Eindhoven. The American end of the operation was a success but the unfortunate British 1st Airborne which dropped at Arnhem found itself confronted by an entire SS Panzer division and, lacking tanks or other heavy equipment, was unable to secure its objective despite a heroic fight. Withdrawn to rest and recuperate, the two American airborne divisions were the only forces in a position to help when Field Marshal Gerd von Rundstedt's Panzer divisions struck unexpectedly through the Ardennes just before Christmas. The 101st raced to secure the road junction at Bastogne, while the 82nd took the one at St Vith, and between them they slowed the German advance until relieved by General George S Patton Jr's Third Army. It was at Bastogne that General Anthony McAuliffe delivered the famous reply 'Nuts!' when called upon to surrender.

Meanwhile, a third airborne division had been created from the First Airborne Task Force, the 17th, commanded first by Major-General Robert T Frederick and then by Major-General William Miley. The Anglo-American Task Force had been formed in Italy for Operation 'Dragoon', the invasion of southern France in August 1944, after which the British component had returned to Italy but the new 17th remained in France. It was rushed to help in the drive to relieve the 101st at Bastogne and subsequently fought its way into Germany after dropping alongside the British 6th Airborne in March 1945 to secure the Rhine crossings.

Finally, on the other side of the world a single regiment, the 503rd, had sailed to Australia after its battles in Tunisia and formed the nucleus of the 11th Airborne Division which fought the Japanese at Leyte, Luzon, Corregidor ('The Rock') and Okinawa under Major-General Joseph M Swing. After the war all the airborne divisions except the 82nd were disbanded although the 101st was subsequently reactivated as an 'Airmobile' division, equipped with helicopters, for the war in Vietnam.

While the US airborne forces were being built up in 1942, a new type of elite unit was also in the process of being created: the Rangers. Their name has its origins in the force formed in 1756 by Major Robert Rogers as a light, fast-moving unit to fight for the then British colonies against the French and Indians. The 1st Battalion of this now illustrious American formation was raised in June 1942 in Northern Ireland. General George C Marshall, impressed by the exploits of British Commandos and anxious that American troops should get a foretaste of battle before any major operations took place, chose a Major from the staff of the 34th Infantry Division to recruit a similar unit from volunteers of any US regiment then in Ulster. William Orlando Darby was delighted for it was just what his experience as a cavalryman and staff officer – as well as his enthusiasm – suited him.

The criteria he looked for were much the same as his predecessors sought in selecting for the commandos and airborne forces: physical fitness and general toughness, courage and self-discipline, good marksmanship and fieldcraft, the ability to swim and map read and, ideally, some experience of mountain climbing and small boat handling as well. Because Darby wanted a good cross-section of experts in ordnance, signals, navigation, mechanical repair and maintenance and other specialist skills, he did not just recruit from the infantry but from assorted other army corps as well, luring them with the promise of adventure and 'first crack at the Hun'.

Unfortunately, the Rangers did not have a happy war. Their first experience of battle was during Operation Torch but they were decimated in Sicily and particularly during the battle to break out of the Anzio beachhead a few months later. They were reduced from five to two battalions. The survivors played a crucial role as pathfinders on D-Day, though, while a new sixth battalion fought in the Pacific, but they were all disbanded at the end of the war – temporarily, as events transpired.

The last of the Anglo-American elite forces, the US Marine Corps, has a long history dating back to its formation by Order of Congress on 10 November 1775 that cannot be recounted here. The Corps fought with distinction from 1941–45, particularly in the 'island hopping' campaign against the Japanese in the Pacific.

Some details of its more recent campaigns and battles are given later. However, mention must be made of the USMC's four wartime parachute battalions. Even though one of them saw no action at all and the other three did not make a single combat drop, they saw heavy fighting on Guadalcanal, Bougainville and Choiseul and are included here because they were the precursors of today's 'Recons' – Marine Reconnaissance Units – who are also parachute-trained and form another elite within an existing elite.

Specific mention should also be made of the First Special Service Force, even though it bears no relationship to the modern US Army Special Forces – the 'Green Berets' – which are a postwar phenomenon. This was a joint US-Canadian unit raised in June 1942 from backwoodsmen and trackers under Lieutenant-Colonel Robert Frederick (later commander of the 17th Airborne Division) and tasked with sabotage operations in mountain and arctic conditions although it was never properly deployed in this role. Nicknamed the 'Devil's Brigade', it saw action in Italy and southern France during 1943–44 before being disbanded. The experience it gained was particularly useful when Canada began organizing its current Special Service Force in 1968.

Since the sweeping German Blitzkrieg of 1940 left the occupied nations with little opportunity to form similar elite units for special missions, and we have already mentioned the Free French and Belgian units in the SAS, it just remains to look briefly at the Soviet Union. Russia, as we saw at the beginning, was an early exponent of airborne warfare and today has the largest paratroop force in the world.

The first Russian parachute unit was formed in Leningrad in 1931 and by 1933 had been expanded to the size of a brigade with gliderborne troops as well as paras. The first fullscale parachute drop in history – other than in exercises – occurred in November 1939 at the beginning of the Winter War with Finland. Even though it was not particularly successful, further expansion followed so that by the end of 1940 there were six full brigades. Despite an acute shortage of suitable aircraft, when Germany invaded the Soviet Union in June 1941 five European corps were in the process of being formed, each of three brigades, with a further brigade stationed in the Far East. The principal use made of parachute forces during what the Russians call the 'Great Patriotic War' was in support of partisan brigades behind the German lines. Although generally poorly planned and coordinated, these attacks tied down large numbers of German security troops guarding supply centres and escorting convoys. The two large-scale airborne operations – near Vyazma in January 1942 and Kanev in September 1943 – were total disasters with almost the entire air landing force being wiped out in each case. For the remainder of the war the airborne brigades acted as ordinary infantry. Their lack of success can be attributed more to poor equipment, training and leadership than to lack of courage or conviction among the troops, 196 of whom were awarded the Hero of the Soviet Union medal.

Finally we come to the Soviet Naval Infantry, equivalent of the British Royal Marines and US Marine Corps. These have a long history dating back to the days of Tsar Peter the Great and during the Second World War were expanded to a staggering 350,000 men organized in 40 brigades and a number of smaller units. They carried out numerous small-scale landings behind German lines and four full-scale amphibious assaults during 1943–44, two on the Kerch peninsula, one at Novorossiysk on the Black Sea and one at Moon Sound in the Baltic. Generally the Naval Infantry were also employed as ordinary infantry, though, and they were disbanded in 1947 – though only for a short time.

From these varied beginnings have emerged the elite forces of the world as we know them today.

The NATO Countries

UNITED KINGDOM

The United Kingdom is a small country with a limited military budget which never stretches far enough, but its soldiers – all volunteers, since National Service was abolished in December 1960 – are widely respected for their professionalism. It often surprises people when they count up and realize that British soldiers, sailors and airmen have fought in more than 50 campaigns since the end of the Second World War, one of which has been running for more than 20 years in Northern Ireland, of course. But despite other obligations in places as far flung as Hong Kong and Belize, the British armed forces are principally dedicated to the service of the North Atlantic Treaty Organisation (NATO), and their main theatre of operations if a confrontation ever occurred with the Warsaw Pact would be in Europe.

For this reason the bulk of the front-line regiments and squadrons are normally stationed in West Germany while the Royal Navy's main role is in the North Sea and North Atlantic. However, their ability to respond quickly and flexibly anywhere in the world was clearly shown by their rapid deployment to the Falklands in 1982, while a naval task force was also deployed to protect international shipping during the 1980–88 Gulf War between Iran and Iraq. In recent years the British armed forces have, inevitably, also been deeply involved in the war against international terrorism, and in this role the Special Air Service Regiment is acknowledged to reign supreme.

The Special Air Service Regiments

It is invidious to make distinctions between the many fine, fit, brave and skilled officers and men in the British or any other army, but the three SAS Regiments demand the highest standards of all and while those volunteers who fail to pass their rigorous selection tests may feel disappointed, none can feel ashamed because to be accepted for the course is in itself an

accolade. The SAS – or 'Sass' as it is pronounced in the army, without the definite article – nevertheless managed to remain largely out of the public eye until 11 minutes in May 1980 focused the world's attention on the force and some of its more esoteric abilities.

On 5 May, concealed television cameras revealed 'live' to an enthralled audience a situation which would have done credit to a thriller film and which has, of course, been imitated in fiction many times subsequently. On the balcony at the front of the white, colonnaded façade of the Iranian Embassy in Princes Gate, London, suddenly materialized the sinister figures of a number of black-garbed, masked and hooded men carrying automatic weapons and grenade launchers. Since the previous Wednesday, 24 men and women including three Britons, had been held hostage by masked gunmen demanding the release of 91 political prisoners held in Iran by the fanatical Ayatollah Khomeini's regime. Quite how they expected to achieve their aims remains a puzzle because the British government had no influence over that of Iran at the time – almost the reverse in fact!

The police stood by the usual 'rules of engagement' in hostage situations, which state that military action may not be taken until all other avenues of discussion have proved cul de sacs or until the terrorists actually kill one or more of their victims. The SAS, although on standby alert, could not be involved other than in an advisory capacity until one of these criteria was met. However, they did plant microphones, dropping them on cables down chimneys, so they could accurately locate where the hostages were being held and where each terrorist was stationed. Any noise the team from the 22nd Special Air Service Regiment (22 SAS) made while on the roof of the building was masked by the sound of pneumatic drills from nearby 'roadworks'.

Tension built. The terrorists, led by a man only identified as 'Oan', had demanded a meeting

Heckler & Koch MP5

The principal German military smallarms manufacturer since the Second World War, Heckler & Koch, was founded in 1948 and began making the army's G3 rifle in 1956 shortly after the Federal Republic was admitted to NATO. This caused a row at the time because the rifle was an improved copy of the CETME design developed from war-time plans by Nazi engineers who had sought refuge from Allied justice in General Franco's Spain! H&K subsequently developed the MP5 (MP derives from Maschinen-Pistole) sub-machine gun which uses many components of the G3 and the heavier HK21 machine gun, saving production costs.

MP5SD1 which does not have a stock at all; the MP5SD2 which has a fixed plastic butt; and the MP5SD3 which has the metal stock. Finally there is a miniaturized version only 325 mm long, the MP5KA1, which is specially designed for covert operations since it can be concealed almost as easily as a pistol. There is no stock, the barrel is almost non-existent and there is a small hand grip just behind and below the muzzle. Range is no more than that of a pistol but the automatic fire capability is useful in counter-terrorist operations.

All MP5 variants are chambered to 9 mm calibre (with a smaller version just being introduced) and have straight or curved 15- or 30-

There is a whole family of MP5s, the MP5A2 being the most common. This has a plastic butt in line with the barrel, a pistol grip and the magazine receiver in front of the trigger. The MP5A3 used by the SAS during the embassy siege is identical except that it has a retractable tubular metal stock, reducing the weapon's overall length and making it especially suitable for use in confined spaces. There are three silenced versions: the

round box magazines. A delayed blowback firing mechanism gives a cyclic rate of fire of up to 800 rounds per minute (rpm). A useful feature is a selector switch enabling the weapon to be fired single-shot, in bursts of two to four rounds, or fully automatic. The sights on the basic MP5A2 are calibrated to 440 yards (400 m) but normal effective range is more like 220 yards (200 m), and even less with the silenced versions.

TOP The miniaturized MP5K. **ABOVE** The silenced MP5SD2.

with representatives of other Middle Eastern countries and subsequent safe conduct to a plane at Heathrow Airport. The demands went unanswered since you do not negotiate with terrorists (even if you pretend to): you just wear them down until they finally realize they cannot win. Then one of the prisoners, assistant press attaché Abbas Lavasani, who was a devoted follower of Khomeini, somehow managed to get to a telephone. The terrorists shot him out of hand. Sir David McNee, Commissioner of Police in charge of the negotiations, rushed the senior Iman of the London Central Mosque to the phone to plead for moderation. 'Why should we wait any longer?', the terrorist leader demanded. Sounds of further shots were heard, then a body was dumped outside the embassy's front door. It was assumed that a second hostage had been killed and the decision was taken: send in the SAS. (In fact the firing had been a bluff and the body was actually Lavasani's, but no-one on the outside could have known that at the time.)

The SAS operation had been meticulously rehearsed. Three teams each of four men, were deployed, one entering the front of the building from the adjoining balcony next door, one abseiling down on ropes from the roof to smash through windows in the rear, and the third bursting in through a wall from next door, the bricks having been carefully removed leaving only the facing plaster in place. In order to gain entrance they used frame charges, following through with stun grenades (colloquially known in the army as 'flash bangs') and tear gas. Their principal weapon – a favourite in the SAS and many other counter-terrorist units – was the German Heckler & Koch MP5 sub-machine gun, popularly known as the 'Hockler'. The members of the assault party were dressed in black overalls over kevlar body armour, with black balaclava anti-flash helmets and gasmasks. In the street outside, police marksmen from D11 – the 'Blue Berets' – waited for targets of opportunity,

Masked SAS men on the front balcony of the Iranian Embassy.

Browning Hi-Power automatic pistols at the ready.

At 7.26 pm on that warm early summer evening, reasoning that the terrorists would be less likely to expect an assault while it was still daylight, the SAS teams went into action. As the first pair of troopers reached the balcony at the front, a second pair abseiled down to the first floor balcony at the rear. Then came a hitch. One of the men in the third pair got his rope snagged. This meant that the troopers below him could not use their frame charges without injuring him, so they had to simply kick in the windows, hurling their 'flash bangs' which cause disorientation but do not generally otherwise harm. As the second pair of troopers swung in through the first floor windows at the rear, 'Oan', who had been temporarily distracted by a carefully timed telephone call, ran to the landing and raised his automatic pistol. Police Constable Trevor Lock, a member of the Diplomatic Protection Group who was among the hostages, flung himself at the gunman, distracting his aim until an SAS trooper shot him.

Meanwhile another hostage, BBC sound engineer Simeon Harris, had fled to the front of the building. Throwing back the curtains, he was astounded to see what he thought was a 'frogman' on the balcony. The SAS man urgently motioned him to move back and four seconds later a frame charge exploded, showering the room with glass, and four soldiers dived into the building. Simultaneously, the third team broke through the adjoining wall.

On the second floor above the 15 male hostages (who had been separated from the women) listened in horror to the explosions and gunfire. When the attack had begun there had been only one guard present in the room, but now two more terrorists ran in and began shooting at them. One Iranian official was killed and two more wounded. But then, as if realizing the futility of their actions, the men threw down their guns and attempted to mingle with the hostages.

Police look on as smoke pours from the Embassy after the rescue.

The deception did not save them – the SAS shot to kill. Only one of the six terrorists, who had been guarding the women at the back of the building, was allowed to surrender. The whole operation had taken 11 minutes and seconds later the relieved hostages began emerging, hustled to safety by the soldiers because several small fires had broken out. It was a miniature classic of an operation, watched by astonished millions as it actually happened, and for really the first time the general public started demanding to know who these SAS 'supermen' were.

The present-day SAS, which consists of three regiments, one Regular and two Territorial, came into being in a rather odd fashion. The wartime 1st SAS Brigade was disbanded in October 1945 but within a few months the War Office (now the Ministry of Defence) decided that there would, after all, be a need for such a force in the postwar world. They reactivated a Territorial Army regiment, 'The Artists' Rifles', as the 21st Special Air Service Regiment (Artists) – (Volunteers) in 1946. Then, in 1951, three years after the start of the Malayan Emergency, Brigadier Michael Calvert arrived. The former CO of 1st SAS Brigade 1944–45, he quickly formed a new volunteer counter-insurgency unit called the Malayan Scouts (SAS). This rapidly grew to a strength of three battalions (which the SAS calls squadrons), one composed of volunteers from other units, one from 21 SAS and one from Rhodesian volunteers. A year later it was re-designated 22nd Special Air Service Regiment and became an established part of the regular army. In 1956 the Rhodesians returned home and were replaced by a New Zealand squadron which included many Maoris and Fijians, all tough fighters. Subsequently a second TA regiment was raised, 23 SAS, broadly based in the north of England while 21 covers the south.

The situation in Malaya was complex and needed careful handling. After the war many local guerrillas who had fought against the Japanese started a campaign to try to establish a communist regime. Operating from camps deep in the jungle, between 1948 and 1950 they had killed more than 1,300 people and the British infantry battalions stationed in Singapore had made precious little headway against them. Calvert inaugurated what was then a revolutionary 'hearts and minds' campaign to win over the local population and deny the guerrillas their support. The SAS men, alongside Gurkhas and Royal Marines, not only fought the communists on their home ground, but also built roads, bridges, clinics and schools while education officers fluent in local dialects preached the principles of democracy. Victory did not come overnight – the emergency lasted until 1960 – but the campaign was eventually successful and has become a model for subsequent operations of similar nature. What might have happened in Indo-China had the French and, later, the Americans tried harder to adopt similar tactics remains speculation.

The task accomplished and many useful lessons learned, not least the techniques of parachuting into treetops and then abseiling on ropes down to the jungle floor, 22 SAS next found itself embroiled in a very different campaign in the pro-British Persian Gulf Sultanate of Muscat and Oman. Here, in the mid-1950s, dissidents supported by the powerful state of Saudi Arabia had been stirring up trouble which the Omani's own Trucial Scouts were unable to suppress. In 1958 the SAS were sent in, and launched a series of surprise attacks on the rebel stronghold on the Jabal Akhdar plateau and in the surrounding mountains, and in three months had suppressed the rebellion. They would return later, but in the meantime trouble had again broken out in the Far East, this time in Borneo.

Although Gurkhas bore the brunt of the fighting during what has subsequently come to be called the 'Brunei Confrontation', in which President Sukarno of Indonesia tried by force to unite Brunei and Sarawak into a new 'communist' state, the SAS was involved from 1963–65

when a coup deposed Sukarno, known to the troops as 'the mad doctor'. Similar tactics to those employed in Malaya worked particularly well with a population which did not want to be ruled from Peking (now Beijing). For the SAS it was now back to the Middle East, first to Aden (now North and South Yemen), then Oman once more in the early 1970s when a mere 10 troopers routed a force of more than 250 rebels at the battle of Mirbat. While in Oman this time the SAS recruited and trained a potent force of national levies, the Firqats, fierce local tribesmen who took with delight to the form of irregular warfare in which the SAS excels and who, by the end of 1975, had brought the country back under the Sultan's control.

In the interim the SAS had become heavily involved in the British army's running battle with those who are fighting for a united Irish Roman Catholic republic. Unlike other army regiments on tour in the Province, who operate openly in the streets, in uniform, the SAS operates clandestinely. Its frequently long-haired personnel, wearing donkey jackets and wellingtons and speaking a perfect dialect, form the 'eyes and ears' of the Intelligence Corps, listening quietly in pubs and feeding back snippets of information which, when pieced together with others, can reveal the hideout of a wanted terrorist or a cache of hidden arms and explosives. This is dangerous, nerve-shredding work with death or mutilation never further away than a single wrong word, and details will remain under wraps for decades – indeed, some will almost certainly never be made public.

The SAS also operates as part of the border patrol network, although the government did not admit this until 1974, five years after the current 'Troubles' began. Teams of four men – the SAS's normal tactical unit – may lie hidden in ambush for days, hardly moving in well-camouflaged hides, enduring filth, damp, cold and meagre uncooked rations until a man or group of men attempt to slip across the border.

Then they strike, and the incident reaches the newspapers a few days later as a random interception by a border patrol from a totally different regiment. Nor are known terrorists who flee abroad safe from 'Sass' as the much-publicized executions in Gibraltar proved. The SAS hates such publicity, for if 'Who Dares Wins' is their motto, secrecy is their watchword. Unfortunately, but inevitably, the regiment came into the public eye again during the Falklands' campaign.

At 4.15 in the morning of 15 May 1982, Griff Evans, a sheep farmer in the small community on Pebble Island (off the coast of West Falkland), was woken by the sound of explosions. Peering through a window he saw the whole night sky brilliantly illuminated by exploding flares, ammunition and oil drums from the nearby airstrip which the occupying Argentine troops were using as an alternative to Port Stanley. As Griff made his anxious wife a cup of coffee, the couple were completely unaware that their house was screened against Argentine retaliation by a 16-man troop from 'D' Squadron, 22 SAS. A second troop had placed explosive charges in the 11 aircraft on the strip while a third stood by in reserve. Then the destroyer HMS Glamorgan brought her guns into play, her shots zeroed in by an artillery observer accompanying the SAS teams. The result was the destruction of all the Argentine aircraft and the radar installation which could have been used to help guide aircraft attacking the British Task Force. It was the first indication that Argentine commander General Menendez had that British forces had arrived in the Falklands, although he was obviously aware that the more easterly island of South Georgia had already been recaptured, along with the submarine Santa Fe.

The men of 'D' and 'G' Squadrons played a major part in the liberation of the Falklands, providing reconnaissance reports for the Commandos and Paras who landed after them and infiltrating Argentine positions at night to kill

SAS trooper, his eyes obscured by the censor, during an exercise in the Brecon Beacons.

sentries, reducing the opposing troops' already shaky morale. Inevitably, they took casualties too, 19 men from 'D' Troop being killed in a helicopter crash and the popular commander of Mountain Troop, Captain John Hamilton, dying while giving covering fire for the other three men in his patrol, an act of valour for which he was awarded a posthumous Victoria Cross. Other SAS teams were reportedly active in mainland Argentina, attacking airfields, although this has never been confirmed. The same is true of persistent reports that the SAS were involved in Afghanistan where their task would have been observation and evaluation of Soviet tactics, particularly those of Russia's own elite Spetsnaz forces. Such operations must, of necessity, remain secret for many years, but over the last decade many hitherto unknown details of SAS recruitment, training and organization have been made public and it is worth looking at these in depth because they serve as a model for other elite forces around the world.

Unlike the elite forces of some nations, the SAS does not recruit directly from the public but only from within the existing regiments and corps of the British army, particularly from the Parachute Regiment and the Brigade of Guards which are already elite forces in their own right. Even then, not all those who apply are admitted to the rigorous selection course, held twice a year, because the SAS looks for a particular type of person. There is no disgrace in having an application refused because a man can be a perfectly good soldier, competent, skilled and reliable, but just lack that psychological make-up which the SAS needs. A primary requisite is self-confidence, but not cockiness or an insubordinate attitude. Coupled to this there has to be drive, determination to excel and, in many ways even more important, to endure in the face of physical hardship, pain and severe mental stress. Physical and mental strength and discipline are thus essential. The initial selection procedure is designed to test these qualities to

breaking point for in an operational four-man team, each individual has to have total, uncompromising faith in his 'mates'.

The selection course, originally designed by Malayan veteran John Woodhouse in the 1950s, is constantly being revised but typically lasts for between three and four weeks. The first half emphasizes physical fitness and map reading with long marches in small groups through the rugged Brecon mountains of Wales. This is designed to bring each recruit up to a common standard. Once the men – their numbers already whittled down by either voluntary or compulsory withdrawal from the course – reach the necessary standard, and have proved that they can stand the pace and navigate accurately in fog or at night using just a map and a compass, they are broken up into smaller groups for even more intensive practice, and are finally sent out individually to complete set tasks within a time limit. A 'sickener' to test their stamina is always thrown in at some point, the recruit arriving at his destination only to be told he has to return to his starting point. This finishes off many aspirants even when they know beforehand that they will have to face the test at some point. The end of the course is 'celebrated' by a gruelling 40 mile (64 km) hike over the mountains carrying a rifle and 55 lb (25 kg) bergen rucksack, an ordeal which has to be completed in 20 hours. Even at the end of all this the surviving 10 to 20 per cent of the original intake are still not members of the SAS, but have to undergo a further 14 weeks of more specialized training.

If this initial emphasis on cross-country marching and navigation seems strange, it has a very logical purpose. The principal task of the SAS in time of war is reconnaissance and sabotage deep behind enemy lines, and each man has to be capable of fending for himself during the approach to the objective and subsequent withdrawal, which could conceivably last weeks. Under such circumstances there is always, of course, the possibility of capture and for this reason the men are taught to memorize map references rather than write them down, and to fold their maps carefully so as to avoid revealing the locality they are interested in, or where they have come from.

The greatly reduced intake of new recruits now spends seven weeks in what is called 'general' training in the SAS's standard operating procedures (SOPs). They learn how to work together in four-man teams, practising reconnaissance, sabotage and ambush techniques and honing the basic skills of first aid, fieldcraft and marksmanship which they will have already learned in their parent regiment or corps. They are also taught how to resist interrogation under very realistic conditions which stop only just short of actual physical torture. Then follow three weeks of combat survival training in which the men learn how to live off the land. This goes

SAS men in action with FN 7.62 mm rifles (below) and (opposite) displaying some of their equipment including high altitude parachute, scuba, mountain, arctic and tropical kit.

much deeper than finding out how to snare a rabbit or 'tickle' a fish, for survival in the field does not just involve eating and drinking, it depends on avoiding detection by the enemy, so the careful siting and camouflage of a 'hide' is just as important. At the end of this phase of training recruits are sent out into some region of rugged countryside equipped with just a knife and a box of matches and have to survive for five days while avoiding detection by the instructors hunting them down. This is part of the 'escape and evasion' training.

Finally comes what most recruits consider the 'fun' part of the course. It is also the most expensive, which is why the army leaves it until last, saving unnecesary costs by ensuring that the surviving recruits are almost certain to make the grade. This is parachute training at Brize Norton, the Royal Air Force station in Oxfordshire.

Trained members of the Parachute Regiment do not have to undergo the basic part of the course but instead help the instructors, although they do have to take the seven jumps – three of them with full equipment and one of them at night – which are the final criteria for admission to 'the Regiment'. (As mentioned earlier, there are in fact three regiments, 22 SAS composed of Regulars and 21 and 23 SAS which are part of the Territorial Army; volunteers for the latter two regiments have exactly the same training as the Regulars but it is spread over a period of up to three years because it has to be fitted in with the demands of the men's civilian jobs. The TA has long since lived down its 'Dad's Army' image and its men are just as tough and professional as the Regulars, even though they often have to make do with handed-down vehicles and weapons.)

For men of any of the regiments, now, at last, comes award of the coveted beige beret with its famous winged dagger badge, but even so they are still only on probation and will remain so for a further nine months (longer in the TA) while they learn more specialized skills and decide, depending on aptitude, inclination and ability, which type of Troop they will finally join.

Each of the three SAS regiments divided into four squadrons, known as 'Sabre' squadrons and lettered 'A', 'B', 'D' and 'G'. In addition, 22 and 23 SAS both have separate signals squadrons. A regiment averages about 600-plus men including headquarters and supporting services and each Sabre squadron is divided into four Troops of 16 men. These are Boat, Air, Mountain and Mobility Troops. The men of Boat Troop undergo much the same training as the Royal Marines' Special Boat Squadron and learn canoeing and small boat handling, offshore and inshore navigation, sub-aqua swimming using either scuba or oxygen rebreathing sets, and the techniques of beach reconnaissance and underwater demolition. The men of Air Troop learn advanced parachuting techniques in-

cluding high altitude/low opening (HALO) and high altitude/high opening (HAHO) jumping. Both these techniques allow them to be dropped – normally from an RAC C-130 Hercules – at 30,000 ft (9,000 m) or so several miles from their objective by day or by night.

Mountain Troop men train alongside the Royal Marines' Mountain & Arctic Warfare Cadre in skiing, rock and ice climbing, mountain navigation and high altitude survival. They also learn how to dig snowholes, burrowing into the snow to create artificial caves both for shelter from the wind and concealment from the enemy, breathing by poking a hole through the 'ceiling' with a ski stick. Finally, Mobility Troop learns how to drive, maintain and repair the wide range of vehicles used not just by the British army but by those of other countries as well, and many of its personnel reach international rally driving standards. (All members of the SAS routinely learn to use and maintain the full range of NATO and Warsaw Pact smallarms too, as do the special forces of other nations, of course.)

Training does not stop here, for every man in the SAS is constantly sharpening his expertise. A normal 'tour' with the Regiment is three years after which a man will usually (but not necessarily) return to his parent outfit, thereby spreading SAS skills gradually throughout the whole army. During these three years many men gravitate from Troop to Troop to increase their knowledge and capabilities yet further. Those with a gift for languages may attend the army's language school at Beaconsfield in Buckinghamshire and some of them will later join army intelligence, or 'I' Corps (often known as 'eye-spy corps'!) Some will spend a period with the Army Air Corps learning how to fly fixed-wing aircraft and helicopters. Other men specialize in the counter-terrorist role and learn the techniques of entering and fighting inside houses at the Regiment's special training centre at Hereford, its headquarters. The Regiment also maintains close links with the special forces of other friendly nations, particularly in the USA, Australia and New Zealand but also with the Belgian 1st Parachute Regiment which is descended from the wartime SAS and with West Germany's GSG9, and there is a great deal of cross-pollination of ideas and techniques as well as carousing in the Mess.

The Parachute Regiment

The three SAS regiments maintain particularly close links with The Parachute Regiment from which a large proportion of their personnel are drawn. At the time of writing the Regiment consists of three Regular battalions headquartered at Aldershot – 1, 2 and 3 Para – and three Territorial battalions – 4, 10 and 15 Para. With the end of the Second World War there was an inevitable rundown in the size of Britain's airborne forces, the 1st Airborne Division being dis-

On exercise in West Germany, a light infantryman with a 7.62 mm General Purpose Machine-Gun, popularly known as a 'Gimpy'.

banded in 1945 and the 6th in 1948 leaving just a single brigade which was given the number 16 to commemorate the two earlier formations.

The 16th Independent Parachute Brigade Group, to give it its full title, saw a great deal of action during the 1950s, one squadron operating alongside the SAS in Malaya and others serving in Egypt or in Cyprus in the battle against EOKA terrorists. In November 1956 the brigade's 3rd Battalion parachuted in to capture the airfield at Port Said in advance of the Anglo-French-Israeli invasion of Egypt designed to retake the Suez Canal which President Nasser had nationalized. Unfortunately, this invasion provoked a worldwide outcry and the occupying forces were compelled by the United Nations to withdraw, leading indirectly to two later wars between Egypt and Israel. The brigade was also involved in supporting King Hussein of

Jordan when trouble erupted in Lebanon and Iraq in 1958, as well as in Aden, Borneo, British Guiana and Anguilla. Inevitably, the brigade was involved in peacekeeping duties in Northern Ireland from 1969 onwards.

On 30 January 1972 an incident occurred which gave the Paras much unwanted and unjustified publicity. The 1st Battalion was on duty in Londonderry tasked with controlling an illegal protest march. A group of youths began bombarding them with stones and CS riot gas canisters, then snipers from the Provisional wing of the Irish Republican Army (PIRA, generally just known by the army as 'Provos') began shooting at them from a block of flats. The Paras returned fire and 13 men in the crowd were killed. The IRA accused them of firing indiscriminately at 'innocent women and children' and the incident has subsequently entered history as 'Bloody Sunday'. In fact, of course, if the Paras had indeed fired indiscriminately the carnage from their 7.62 mm self-loading rifles (SLRs), each with a 20-round magazine, would have been far higher. The IRA exacted revenge on 27 August 1979, ambushing a convoy of trucks at Warrenpoint and killing 16 of the regiment's men, an incident overshadowed by the assassination of Lord Louis Mountbatten on the same day.

By this time 16 Parachute Brigade no longer existed, having been disbanded by a cost-conscious government on 31 March 1977 and all that was left was a single airborne battalion in what was then known as 6 Field Force, the other two battalions serving purely in an ordinary infantry role. Each battalion consists of six companies: headquarters, support and four rifle. The army fortunately maintained airborne training by rotating the three battalions and when Margaret Thatcher became Prime Minister in 1979 one of the first things she did was authorize two airborne battalions, 2 and 3 Para, with all necessary aircraft and equipment. These formed part of 5 Infantry Brigade from the beginning of 1982 while 1 Para was back on duty in Northern

OPPOSITE British special forces have had some of their most notable successes in jungles from Malaya to Belize, and are seen here with a 'Gimpy' on its sustained fire tripod.

Ireland – to their great chagrin for it was 2 and 3 Para, seconded to 3 Commando Brigade, Royal Marines, who reaped glory in the Falklands.

Lacking at the time an equivalent of the American Rapid Deployment Force, it was a tribute to the British armed forces, to the mercantile marine and to the resolution of the government that the Task Force was on its way to the South Altantic within seven days of the Argentine invasion on 2 April 1982. 3 Para, on leave for Easter and with one of its officers on his honeymoon, was recalled in record time and sailed aboard the liner SS Canberra on the 9th, followed by 2 Para and all support teams and equipment on the SS Norland on the 26th. It was actually 2 Para which had the distinction of being the first major unit ashore at San Carlos on 21 May, although SAS and SBS reconnaissance teams had been operating on East Falkland for some time.

The SAS had reported an Argentine force of approximately battalion strength occupying Darwin and Goose Green settlements to the south of San Carlos and the Task Force commander, Rear Admiral 'Sandy' Woodward, ordered 2 Para to take this out so as to secure the ground forces' right flank. The battalion moved off across the mountains and was in position on 27 May. Unknown to them, though, the Argentine garrison had been reinforced by two more battalions so the Paras would be attacking at odds of one to three – the exact reverse of the ratio normally recommended for an assault on an enemy in prepared positions. Moreover, although 2 Para had an unusually high quota of 7.62 mm General Purpose Machine Guns ('Gimpys') because it had been preparing for a tour in Belize at the time of the Argentine invasion, its only support weapons were a mere three 105 mm Light Guns and two 81 mm mortars.

The battalion moved off from its start line at 3 am on the 28th, led by its commander, Lieutenant-Colonel 'H' Jones, and immediately ran

ABOVE A candidate for the Parachute Regiment slogs his way through the arduous assault course.

OPPOSITE They have to be prepared for all situations, and all climates.

into stiff opposition from well-sited .50 calibre heavy machine-guns and artillery. Nevertheless, by midday the Paras were in a strong position although 'A' Company was pinned down by a couple of strong machine-gun positions. It was this point that 'H' led a flanking attack to eliminate the obstacles; he was hit several times and died shortly afterwards, but the Argentine positions were overrun. Colonel Jones was subsequently awarded the Victoria Cross for an action very much up to British airborne forces' tradition, although it was subsequently questioned whether it was the proper role of a battalion CO to lead such an assault in person. Command was assumed by Major Chris Keeble and by nightfall 2 Para had taken Darwin and surrounded Goose Green despite an air attack by four Argentine aircraft. Keeble, worried about potential casualties among the civilian population if the Paras launched a direct assault, sent in a prisoner with a surrender demand in the morning and two Argentine NCOs emerged with a white flag. A Spanish-speaking Royal Marines' officer, Captain Rod Bell, accompanied them back into Goose Green to meet the garrison commander, Air Vice-Commodore Wilson Pedrozo. At one o'clock the Argentines laid down their arms and lined up to surrender. The Paras were astounded – 'gobsmacked' in their own words – to find themselves in charge of 1,350 prisoners plus 190 dead and wounded. 2 Para's own casualties were 15 dead and 30 wounded. It was a truly remarkable feat of arms.

Meanwhile, Lieutenant-Colonel Hew Pike's 3 Para had been force-marching overland towards Teal Inlet on the northern coast, which they reached on the 29th. They then moved up to Mount Kent, to the west of the islands' capital, Port Stanley. Here they dug in to await the arrival of artillery support, which was delayed by an acute shortage of helicopters. Their next objective was Mount Longdon, which was heavily defended by the Argentine 7th Infantry Regiment and a number of crack Buzo Tactico troops

– Argentine equivalent of the SAS. The attack finally went in during the night of 11/12 June and it was immediately apparent that the Argentine forces here were not going to give in without a tough fight. Explosions from grenades, rockets and artillery fire rent the night air while deceptively lazy streams of tracer from dozens of machine guns flew through the moonlit sky. A minefield blocked one avenue of approach and the sheet weight of fire made it difficult to get near the well dug-in machine gun nests. One in particular held up the advance until Sergeant Ian McKay of 'B' Company led four of his men in a frontal assault, charging up the hill with blazing SLRs. McKay seemed to lead a charmed life as one after another of his men fell, but he was finally hit just as he reached the parapet. Although he was also killed, his falling body blocked the breech of the machine gun and the rest of his Company was able to capture the position. Like 'H' Jones and John Hamilton, Ian McKay was awarded a well-deserved posthumous Victoria Cross.

With Mount Longdon secured, it was 2 Para's turn again the next night. The objective was Wireless Ridge, one of the last remaining pieces of high ground overlooking Port Stanley. The battalion had been flown in by helicopter following its victory at Goose Green and this time its assault was fully supported by artillery and the light Scorpion and Scimitar tanks of the Blues and Royals. As one observer commented, 'the Paras loved it' as the heavy firepower dislodged one Argentine position after another. Only three of the battalion's men fell during the final assault and when daybreak arrived on 14 June, Keeble's triumphant troops had the literal drop on the disorganized and demoralized Argentine garrison in Port Stanley. Jauntily, wearing their berets instead of their helmets, they sauntered, strolled or ran down the hill, many of them yelling at the tops of their voices, past despondent Argentine soldiers who did not even put up a token resistance. General Menendez formally surrendered later in the afternoon.

In October the following year 5 Infantry Brigade was rechristened 5 Airborne Brigade and entrusted to the command of Brigadier Tony Jeapes, a former CO of 22 SAS. The Falklands' experience had shown the need for a British air-trained rapid deployment force and the brigade currently consists of Nos 2 and 3 Para, No 2 Battalion King Edward VII's Own Gurkha Rifles (The Sirmoor Rifles), a squadron of Fox armoured cars and Scorpion and Scimitar light tanks from the Life Guards, No 7 Field Regiment, Royal Horse Artillery (RHA), No 9 Para Field Squadron, Royal Engineers (RE), Nos 20 and 50 Field Squadrons, RE, No 61 Field Support Squadron, RE, No 23 Field Ambulance, Royal Army Medical Corps (RAMC), No 63 Squadron, Royal Corps of Transport (RCT), No 10 Field Workshop, Royal Electrical and Mechanical Engineers (REME) and No 82 Ordnance Company, Royal Army Ordnance Corps (RAOC). Airlift capability is provided by Nos 47 and 70 Squadrons, RAF, based at Lyneham in Wiltshire, both of which played key roles in providing aid after the Mexico City earthquake in 1985 and during the Ethiopian famine with their C-130 Hercules aircraft. Finally, helicopter support is provided by the Gazelles of 658 Squadron, Army Air Corps.

The modern brigade is therefore a go anywhere, anytime force, ready to respond at short notice to an emergency in any part of the world. No 1 Para is not part of the brigade but instead forms a component in the Allied Command Europe Mobile Force (AMF), a multinational task force specifically trained in arctic warfare and dedicated to demonstrate NATO solidarity in the case of an invasion of Norway.

Parachute training after the arduous initial induction period into the Parachute Regiment comes, as in the SAS, almost as a relief because each man knows that this is the final hurdle before he can wear his 'wings' and the famous 'red' beret. The whole course lasts 23 weeks,

the first seven of which are principally designed to toughen recruits up physically for what is to come later. Actual strength, in the weightlifting sense, is not required although it can come in handy. Stamina, the ability to keep on going when every fibre of the body screams for rest, is far more important – as the recruits soon discover. These seven weeks culminate in a realistic exercise in the Welsh mountains, after which each recruit's performance is examined by a selection committee. Some men are weeded out at this stage, others are given a second chance to prove their mettle and the remainder

Paras with a Milan anti-tank missile firing post in Norway.

progress to the second stage of the course. For them the next three weeks involve advanced weapons training although the physical toughening-up continues. There is then a second review board before successful candidates go through to the third stage which involves crossing obstacle courses with some stretches over ropes up to 50 ft above the ground to test confidence, compulsory boxing matches to test aggression and cross-country races carrying various loads. At the end of this, successful recruits are allowed to wear the red beret but still have to earn their parachutist's wings.

Following a further four weeks instruction in fieldcraft and tactics, recruits are finally sent to Brize Norton where they are taught the correct way to fall and practise controlled drops from a hangar roof and then the tower before they make their first actual drop from a balloon. The British army is one of the last in the world to retain balloons for parachute training but there is an excellent psychological reason for it is actually more difficult to steel yourself to jump from a stationary platform hundreds of feet above the ground than it is from an aircraft. Some candidates balk at this hurdle and are reluctantly eased out of the army or into a different regiment but the survivors go on to complete seven jumps from a Hercules and are then awarded their wings. Apart from the balloon jump, parachute training in other armies follows a very similar pattern.

Gurkhas through the ages: a 10th Gurkha Rifles machine-gunner from 1930 (below) and (opposite) a junior officer and radio operator from the 6th Queen Elizabeth's Own during a modern jungle exercise.

The Gurkhas

The feisty Nepalese hillmen who have formed part of the British army since 1816 are decidedly an elite, respected by all, liked by their allies and feared by their foes. The presence of the 1st Battalion, 7th Duke of Edinburgh's Own Gurkha Rifles (then part of 5 Infantry Brigade) in the Falklands provoked panic among the Argentine conscripts facing them. The reason was the grossly exaggerated horror stories circulating (helped along by propaganda broadcasts from ships in the British Task Force!) about the uses to which the Gurkhas could put their favoured weapon, the kukri. This thick, curved leaf-shaped knife, which comes in a variety of different sizes, is both a useful chopping tool and a much more effective close-combat weapon than a bayonet. Its cutting edge is kept razor sharp and it can inflict hideous

wounds, with the result that many Argentine soldiers feared mutilation if they fell into Gurkha hands. The Gurkhas, of course, do not practise mutilation, being as highly trained and disciplined as anyone in the British army, but these fears helped spur the Argentine surrender when the battalion moved up alongside the 2nd Battalion, Scots Guards, to take Mount Tumbledown and Mount William overlooking Port Stanley. While the Guards had a tough fight for the former feature, the defenders of the latter simply ran away once they realized who they were facing!

Gurkhas are all volunteers and the young 17-year-olds who flock from their remote villages to the recruiting depots at Pokhara and Dharan each year employ all sorts of tricks to gain acceptance, including falsifying their ages and drinking several pints of water to put on weight before

Gurkhas have a fearsome reputation, but their professionalism is also renowned. ABOVE Map reading exercises are an essential part of navigation anywhere, but especially in the jungle. OPPOSITE The Sterling sub-machine-gun, being shorter than an ordinary rifle, is ideal for the close-quarters work that jungle warfare entails.

their medical examination. Nevertheless, there are many disappointed faces and even tears at the end of the initial selection, for the six battalions in the Brigade of Gurkhas – now known as Gurkha Field Force – can only accept approximately one in 15 of the young hopefuls. Many, rejected first time round, try again the following year, while others opt to join the Indian army.

Apart from height and weight, physical stamina, general intelligence; literacy and numeracy are among the criteria for acceptance, for service in the Brigade is regarded as a privilege rather than a right. Apart from the obvious incentives of decent pay and an opportunity for travel which would otherwise be out of the question, the youngsters are motivated by the glamorized stories their elders tell of action in North Africa and Italy during the Second World War or of Malaya, Borneo and other campaigns sub-

sequently, for Gurkhas have fought just about everywhere the British army has been this century.

The jubilant few accepted at the end of each induction period are flown to Hong Kong and established in the depot at Sek Kong in the New Territories where they will spend the next 32 weeks undergoing basic training. This is longer than normal because the recruits have to learn about the outside world as well as eating heartily to build up their weight and strength and learning codes of conduct and discipline totally foreign to them. Those who endure this period – and there are very few voluntary dropouts – go on for a further eight weeks' training in small-arms use, fieldcraft, navigation and, in particular, internal security duties because the Brigade's principal task in peacetime is the defence of Hong Kong against smugglers and illegal immigrants. The initial period of service is four years (and a soldier does not get home leave until he has served three) but there is fierce competition for promotion to NCO status and the opportunity to stay on longer, since 15 years is the minimum qualifying period for a pension.

From a quarter of a million men in 55 battalions at the end of the Second World War the Gurkhas had shrunk to ten regiments by the time India

ABOVE Gurkhas do not just train for jungle warfare – these Canadian prairies could just as easily be Asiatic steppes.

BELOW A night reconnaissance patrol in the New Territories of Hong Kong.

and Pakistan gained independence in 1947, after which four regiments remained part of the British army and the remainder joined the Indian army. The current line-up is six battalions in four understrength regiments, two each in the 2nd King Edward's Own and the 7th Duke of Edinburgh's Own and one apiece in the 6th Queen Elizabeth's Own and 10th Princess Mary's Own. In addition the Gurkhas have their own Signals, Engineer and Transport Regiments, making the Brigade completely self-contained. The battalions are rotated on a regular basis, one being stationed at Church Cookham as part of 5 Airborne Brigade and frequently providing guards and a band for ceremonial duties

at Buckingham Palace. While stationed in the UK, many Gurkhas also take advantage of the parachute training offered at Brize Norton although there is no longer an independent Gurkha parachute battalion. The others are based in Hong Kong, four active, one training and one seconded to Brunei. Gurkhas have also seen action in recent years in Belize as well as the Falklands. Apart from their kukris and smart brimmed hats or ceremonial 'pillboxes', they are uniformed and equipped just like the rest of the British army. Despite their small numbers – currently just above 8,000 – they remain one of the most formidable fighting forces in the world, particularly in jungle warfare.

Gurkhas patrolling an inland waterway in Malaysia – hunting for drugs and illegal immigrants are amongst their most important tasks.

The Royal Marine Commandos, Special Boat Squadron and Mountain & Arctic Warfare Cadre

While the army Commando battalions were disbanded shortly after the end of the Second World War, three of the Royal Marine units were retained – Nos 40, 42 and 45 Commandos – plus a number of smaller specialized units. No 41 Commando was revived to take part in the Korean War 1950–52, disbanded, reformed in 1961 and finally disbanded for the third time in 1968. Subsequently the three Commandos, each approximately 650 men strong, have been involved in Malaya, Borneo, Uganda, Kenya, Tanganyika, Egypt, Cyprus and, of course, Northern Ireland and the Falklands. In fact, it was Royal Marines who saw first action against the Argentine invaders when they landed on East Falkland on 2 April 1982. Present on the island was a tiny force of 92 Marines and Royal Naval personnel from the ice patrol vessel *HMS Endurance*. They miraculously succeeded in beating off the first attack which was spearheaded by 150 members of the Buzo Tactico, but when the Argentines started landing LVTP-7 Amtrack amphibious armoured personnel

carriers the islands' Governor, Rex Hunt, ordered them to lay down their arms to prevent civilian casualties.

There was also a small party of two dozen Royal Marines on the remote island of South Georgia when the Argentines landed there the following day. They succeeded in disabling two helicopters and holing the Argentine corvette *Guerrico* with an anti-tank missile before they were surrounded and forced to surrender. Retribution was swift. Three weeks later a scratch force of 75 Commandos, SAS and Special Boat Squadron personnel landed by helicopter from the destroyer *HMS Antrim* (which provided fire support) and quickly overwhelmed the Argentine garrison which surrendered on 26 April.

Nos 40, 42 and 45 Commando form the principal component of 3 Commando Brigade which at the time of the Falklands' conflict was commanded by Brigadier Julian Thompson. The Brigade is commpletely self-contained and includes two army units, 29 Commando Regiment, Royal Artillery, with 105 mm guns and 59 Independent Commando Squadron, Royal Engineers. It has its own helicopter squadron, 3 CBAS, equipped with Gazelle utility and Lynx anti-tank aircraft, while the Fleet Air Arm provides logistic support and troop-carrying ability with the larger Sea King HC4s of 845 and 846 Squadrons. In addition there is the Commando Logistic Regiment which provides support to 3 Commando Brigade and all other Royal Marine units, 539 Assault Squadron and an Air Defence Troop equipped with Blowpipe anti-aircraft missiles. A Territorial Army unit, 131 Commando Squadron, Royal Engineers, would also serve as part of the Brigade in time of war.

The Brigade's main role in NATO is the defence of northern Norway and all its members are trained skiers, spending three months every winter in Norway alongside 'Whiskey' Company, Royal Netherlands Marine Corps, and the Norwegian army practising arctic warfare. How-

LEFT A member of an SBS beach reconnaissance team stealthily measures the approach route for a raiding party.

ever, it has the flexibility to respond to a threat anywhere in the world and was embarked to the Falklands in record time aboard the liner *SS Canberra*, the Royal Fleet Auxiliary *Stromness* and the carrier *HMS Hermes*. They landed at San Carlos on 21 May; 40 Commando remained in reserve to secure the beachhead against a possible disaster while 42 and 45 Commandos headed for Port Stanley. The loss of the MV Atlantic Conveyor with its cargo of Chinook heli-

Canoeing and scuba diving are essential SBS skills.

copters meant it was only possible to airlift 42 Commando, so 45 Commando had to march (or 'yomp') in company with 3 Para across the whole width of the island to Port Stanley. The going was punishingly hard, the coarse tussocks of grass and soft mud combining with the weight of the men's bergens, weapons and ammunition to produce a number of sprained ankles. One sergeant commented colourfully that the advance looked 'more like the bloody retreat using their Milan anti-tank launchers as 'bunker busters' and, although there were inevitable casualties, both attacks were a complete success and, as we have already seen, the Argentine commander surrendered on the afternoon of the 14th.

The other Royal Marine units active in the Falklands were the Mountain & Arctic Warfare Cadre and the Special Boat Squadron. The M&AW Cadre was originally formed as an instructional unit to train other Marines in mountain leadership, but in 1981 it was given an operational role as a deep arctic and mountain penetration unit for behind the lines reconnaissance and sabotage missions. Its members are all fully qualified Commandos who have already been selected as junior NCOs, so all volunteers for the Cadre are fit, tough and highly skilled from the start. Even so, the seven-day selection course is so arduous that on average less than half of the 'hopefuls' make it. The survivors spend another two weeks learning basic mountaineering skills in Wales, two weeks in Plymouth practising seaborne assaults using small raiding craft at night, then deploy to Norway where Norwegian army instructors teach them how to ski. After a period of Christmas leave they return to Norway for three months' intensive training in arctic survival, a course which culminates as in the SAS with a 40-mile hike carrying full kit across the rugged snow-covered terrain.

Even that is not the end of things for the M&AW Cadre recruits for they next spend several weeks practising navigation and path-finding in Scotland, learning to parachute at Brize Norton, and being taught sniping techniques at Lympstone in Devon. Any member of the Cadre can kill a man at a range of over half a mile with a single shot. More time is spent practising advanced mountaineering techniques in Switzerland and the qualified recruits finally proceed to the Lake District to learn how themselves to be instructors. Graduation is celebrated by an annual month-long exercise code-

from Moscow'! But the Marines made it and by the end of the first week in June the Argentine forces in Port Stanley were surrounded.

Then, while 3 Para launched its assault on Mount Longdon during the night of 13/14 June, 42 Commando attacked Mount Harriet and 45 Commando Two Sisters. The men had been led to believe that the Argentines were all conscripts of low fighting standard and were disconcerted at the strength of the resistance, particularly from well-entrenched heavy machine gun positions. Several men were also mutilated by anti-personnel mines. The Marines retaliated by

M16 Armalite at the ready, and unusually wearing a 'dry' instead of a 'wet' suit, a member of the SBS wades ashore with a waterproof bag of kit.

named 'Ice Flip' in Switzerland. It can be appreciated that members of the Cadre are regarded with the highest respect by the toughest Commando! In the Falklands the Cadre was deployed ahead of the Task Force, alongside the SAS and SBS, reconnoitring enemy strengths and dispositions, and had one short, sharp but successful engagement with Argentine Commandos at Top Malo House, a skirmish which was conducted in such a classical manner that the Cadre later recreated it in a video film which is used for instructional purposes throughout the British armed forces.

Training for membership of the Special Boat Squadron is equally demanding and lasts a full year, but the emphasis is on different skills since the Squadron's primary role is reconnaissance of suitable landing sites for a larger amphibious task force, and this was their main task in the

Trying out a Jet Raider high speed, shallow draught assault boat.

Falklands. As well as basic parachuting they learn the same HALO and HAHO techniques as the SAS and spend many weeks perfecting their canoeing and underwater swimming capabilities. They also learn how to exit from submerged submarines – the SBS team which landed on South Georgia in 1982 was actually flown to the South Atlantic in a Hercules from which they parachuted to a waiting sub and then made the journey ashore in Gemini inflatable boats. Otherwise, training is much the same as in the SAS or M&AW Cadre and the SBS is thus another elite within an elite. The same is true of Comacchio Group, based in Scotland with the principal task of defending Britain's offshore oil and gas rigs against terrorist attack. Scuba diving, parachuting and a high level of marksmanship are obviously demanded of members of this special force as well.

UNITED STATES OF AMERICA

The SEALs

The US Navy's SEAL (Sea-Air-Land) teams are the equivalent of the Special Boat Squadron and SAS Boat Troops, with whom they often interchange to swap information and techniques. They are generally regarded as the most highly skilled and trained fighting men in the American armed forces – though members of the other elite units might quibble with this judgement. Their origins go back to the wartime Underwater Demolition Teams (UDTs) formed in 1943 to assist in the difficult and dangerous task of clearing safe lanes through the German beach defences for the hundreds of landing craft used in the invasion of Normandy. Other UDT teams saw action at Guam, Iwo Jima and Okinawa. Similarly, during the Korean War they prepared

A UDT trainee rolls overboard from an inflatable boat which he will later have to clamber back aboard while it is being towed at high speed.

the way for the amphibious landing at Inchon. UDT personnel are all volunteers from the Navy and Marine Corps and are trained in every aspect of small boat usage as well as underwater swimming and, obviously, learning bomb disposal techniques and the use of a wide variety of explosives.

In 1960 a US Navy study suggested the need for even more specialized units to conduct deep penetration reconnaissance, sabotage and counter-terrorist operations from the sea or rivers, and in 1962 President John F Kennedy gave permission for the establishment of two new teams which were given the appropriate acronym SEALs. Recruited principally from the UDTs, they performed brilliantly during the Vietnam War from 1966 onwards, particularly in the Mekong Delta where they used the numerous rivers to gain access to their targets, sometimes

operating alongside South Vietnamese special forces, the 'Lin Dei Nugei Nghai'. Their missions included reconnaissance, identifying and destroying arms caches and jungle weapons' factories, laying ambushes in 'search and destroy' missions, capturing Vietcong officers for the intelligence information they could provide and even rescuing American servicemen from North Vietnamese prisoner-of-war compounds.

Subsequently the number of SEAL teams has been steadily increased and there are currently six with their Pacific headquarters at Coronado naval base, San Diego, California, and Atlantic headquarters at Norfolk, Virginia. Of these, No 6 is the specialist counter-terrorist unit and is solely comprised of proven veterans rather than recent recruits; it works closely with Delta (see page 62) and the British Comacchio Group.

SEALs during an infiltration/exfiltration exercise with a luridly camouflaged Bell UH-1 Huey helicopter.

SEAL detachments are normally also present at Subic Bay in the Philippines, as well as in Italy, Scotland and Puerto Rico. Each full team consists of 27 officers and 156 enlisted men divided into five platoons. Like the members of the SAS and SBS, they are 'go anywhere' soldiers who can operate equally well in temperate, jungle, desert or arctic conditions with only marginal preparations.

SEALs are recruited from volunteers who have already gone through the arduous 24-week UDT course held at Coronado. The first six weeks is a physical toughening-up process with long endurance marches and swims in the ocean. Recruits have to run everywhere and have to endure press-ups and other physical punishments at the drop of a hat. Less than half of each intake, perhaps 40 to 50 men, succeed in completing this initial 'tadpole' course. This

M16 Armalite and variants

The M16 family of firearms is, alongside the Soviet AK-47 quantitatively the most important in the postwar world and is used by the elite forces of most Western nations in one form or another in preference to their own army's standard rifle. Designed by Eugene Stoner for Armalite, the M16 assault rifle is actually manufactured by Colt and was adopted by the American army in 1961. Modifications introduced in 1966 after experience in Vietnam led to the designation M16A1 which is the most widely used variant. This gas-operated rifle fires 5.56 mm rounds from 20- or 30-round box magazines, either single-shot or fully automatic with a rate of fire of up to 150–200 rounds per minute.

However, it was found that firing in the automatic mode resulted in gross waste of ammunition so in 1981 a selector switch was introduced to give a three-round burst capability which is much more economical. Other changes in the M16A2 include a heavier barrel rebored to take the more powerful NATO standard SS109 5.56 mm cartridge which increases the weapon's range from 340 to 550 yards (310 to 500 m). All M16s can be fitted with the M203 40 mm grenade launcher beneath the barrel. This has a separate trigger and can fire a variety of fragmentation or smoke grenades. The M16 can also be fitted with a telescopic sight or a passive light intensifier.

A variant specifically developed for use by US special forces is the M15 Colt Commando. This has a shorter barrel with a prominent flash suppressor and a retractable butt, and is handier to use in the jungle or confined spaces such as buildings during fights against urban guerrillas, but effective range is only just over 200 yards (180 m). As a footnote, since it is often misunderstood, a flash suppressor at the end of a barrel is not designed to conceal the weapon from the enemy, but to help the man firing it from being temporarily blinded.

M16 with M203 grenade launcher.

course also includes a week-long escape and evasion exercise as well as advanced weapons training. The third part of the course involves learning basic hydrography for beach surveys, plus demolition and communications. Recruits then go to Fort Benning for the army's three-week parachute course which is similar to the British one with the exception of balloons.

At this point the paths of those who are going to stay in the UDTs and those who are going into the SEALs partially diverge although they may overlap at later points, the former going on to an intensive 33-week course in the finer points of EOD which includes dealing with chemical, biological and even nuclear weapons. SEALs instead go on a 10-week course in the use of the Navy's Swimmer Delivery Vehicles (SDVs), which they will have learned something about already. These are small open craft propelled by an electric motor, the modern equivalent of the wartime 'human torpedos'. Being made mainly of non-ferrous materials, they are virtually undetectable to radar or sonar. These submersibles carry up to six men and have independent air supplies into which the men can plug to save the air in their backpack cylinders. The men also have to learn how to exit from and return to submerged sub-marines. The US Navy at the time of writing has three boats specifically converted for clandestine operations. Those SEALs who display the most proficiency in these techniques will be assigned to one of the Navy's two SDV teams whose job is to get the other six teams to their objectives.

As in the SAS and other truly elite forces, training never really stops and members of SEAL teams have to master HALO and HAHO techniques with steerable parachutes, study foreign languages and jungle warfare, learn counter-terrorist tactics, signalling, advanced first aid and a variety of other skills including the most sophisticated unarmed combat techniques and the use of the Smith & Wesson Mk 22 silenced pistol. This is made of stainless steel so

phase culminates in 'Hell Week', a seven-day exercise always held under the wettest and coldest conditions possible with the recruits having to carry their inflatable boats everywhere in forced cross-country marches and then paddle them to a beach for a night infiltration exercise. This reduces the 'hopefuls' by about half again.

For the survivors, the next few weeks involve even longer open sea swims and instruction in small boat handling and scuba diving at a variety of depths using different breathing mixtures. Classroom work is principally concerned with explosive ordnance disposal (EOD) but also involves tactics and survival, and this part of the

Members of the Green Berets during a gruelling joint training exercise with US Navy SEALs.

Two scenes from life in the UDTs and SEALs. **LEFT** Recruits unable to keep up during a 4 mile (6 km) run are punished by press-ups on the beach. **ABOVE** The pressure is on again as they deploy an inflatable boat during a readiness inspection.

it will not suffer in seawater and was introduced during the Vietnam War to deal with guard dogs. With somewhat sick humour, it is generally called the 'hush puppy'. Naturally, they also train with other NATO and Warsaw Pact weapons and regularly use shotguns in 'close' combat situations. SEALs are also the only soldiers known to have regularly used the Stoner/Cadillac-Gage model 63 assault rifle in Vietnam. This imaginative weapons system used interchangeable parts to allow it to impersonate anything from a sub-machine gun to a heavy machine gun with tripod sustained-fire mount proved to be a 'jack of all trades and master of none' and was soon dropped.

Although it sounds even more like science fiction, selected members of the SEALs also learn to operate alongside dolphins. The dolphin is not normally an aggressive mammal, although a shark does not stand a chance against one and there have been several instances of dolphins coming to the rescue of humans in trouble

in the sea. About 60 dolphins reliably reported to have been trained to kill on command were used during the Vietnam War but whether they received a Unit Citation is not publicly recorded! In the SEALs (and their dolphins), the US Navy has one of the finest bodies of special forces in the world, and they further proved their worth in 1983 when a team was first ashore on the Caribbean island of Grenada to help in the rescue of imprisoned American medical students, even though the main part of the operation was carried out by the Marines, Rangers and 82nd Airborne Division.

US Marine Corps

The US Marine Corps is the world's largest elite force with a current establishment of nearly 200,000 men and women. During the Vietnam War they took part in almost all the major operations using helicopters or landing craft to reach

ABOVE Pump-action shotguns are favoured close-quarter weapons among many elite forces, including the SEALs.
OPPOSITE US Marines – seen here after being dropped by a Boeing Vertol CH-47 helicopter – have to learn to live and fight under all climatic extremes.

their targets and have subsequently been deployed in Grenada, Beirut and Panama to name but three recent examples. They suffered over 100,000 dead, wounded or missing in Vietnam and 241 killed by a single terrorist bomb attack in Beirut in October 1983, which demonstrates something of the carnage of what has been called 'war in peace'.

Today there are four Marine Divisions and four complementary Air Wings. A division consists of three infantry regiments, each of three battalions, plus an artillery regiment, a tank battalion, an armoured amphibian battalion with LVTP-7 Amtracks, a light armoured assault battalion with M2/M3 Bradley mechanized combat vehicles and various other supporting services. The infantry battalions are smaller today than in the past but still consist of the traditional three rifle companies, one headquarters company and a support company. The total strength of

ABOVE Marine riflemen during an assault demonstration. **LEFT** In a scene enacted many times over the last 50 years, Marines wade ashore from a landing craft.

each division averages 17–18,000 men.

Basic training at one of the two 'boot' camps at San Diego or Parris Island lasts 11 weeks, officer candidates then going on to Quantico. This is designed to instil physical toughness and confidence as well as smartness, discipline and marksmanship. Those who pass go on to learn more specialist skills according to aptitude, learning how to board and exit landing craft and helicopters. The Marine Corps needs specialist signallers, drivers, combat engineers, medics and men in a dozen and more skilled trades but only the top of each intake can volunteer for the Marines' own elite, the Reconnaissance Units or 'Recons'.

The idea for the Recons was devised by Lieutenant-General Bernard Trainor who served as an exchange officer with the British Royal Marines in 1958–59 and was impressed by the Commando operations in Cyprus. On his return to the United States he pressed for the creation within the USMC of small commando-style units to operate behind enemy lines on reconnaissance missions during which they would, if necessary, remain hidden in one spot for days at a time observing enemy troop movements and reporting back by radio. The first teams were established by 1961 and saw extensive service in Vietnam, usually operating in seven-man squads on 'Sting Ray' missions, plotting Vietcong positions and either calling in artillery or air strikes or taking them out in ambushes. Later, the strength of the squads was reduced to four men as in the SAS or SBS.

Training for the Recon teams is very similar to that of the SBS and includes swimming, scuba diving, canoeing, parachuting, forward artillery observation, beach reconnaissance and demolition work. Today, each Marine division has a Recon battalion of about 500 men. Competition to get into the Recons is keen but as in the British armed forces, only the very best pass the arduous course to win the green and gold winged parachute badge with the legend 'USMC RECON'.

Finally, the four Marine Air Wings consist of between 18 and 21 squadrons with 286–315 fixed-wing aircraft and helicopters. These include F-4 Phantom and F-18 Hornet fighter/ attack aircraft, A-4 Skyhawk, A-6 Intruder and AV-8A/B Harrier attack aircraft, KC-130 in-flight refuelling tankers and a mixture of AH-1, CH-35 and -46 and UH-1 utility, reconnaissance and attack helicopters.

The Rangers

The US Army's wartime Ranger battalions formed by Orlando Darby were disbanded after the Korean War although a cadre remained at Fort Benning as a leadership school to train the army in light infantry tactics. They were not activated during the Vietnam War but in 1975 it was decided to create two new battalions. These were placed on standby alert during the Tehran Embassy crisis in 1980 but were stood down after the failure of the operation 'Eagle Claw' rescue mission (see page 63) and did not see action until they were sent in to Grenada in 1983 where they put up a creditable performance against the Cuban troops on the island.

In 1984 a third battalion was raised and the current status of the 75th Ranger Regiment, to give it its full title, is a regimental headquarters

OPPOSITE Marines storm ashore from LVTP-7 Amtrack amphibious armoured personnel carriers while Sikorsky S-58 helicopters hover overhead. **BELOW** On the beachhead, a battery of M102 105 mm howitzers goes into action.

of some 130 men and three light infantry battalions of about 575. Volunteers for the Rangers will normally have gone through the Ranger School before applying, and have to be parachute-trained before they can be accepted. The school is a unique institution because of its survival when there were no active Ranger battalions. The course there lasts 58 days and teaches volunteers the skills of navigation, survival, advanced weapons handling, close combat with knives or bare hands, and mountaineering.

Those who have passed this course, and completed at least eight parachute jumps, have to go through a three-week selection course called the Ranger Indoctrination Program. This stresses physical prowess but embraces all aspects of fieldcraft and weapons' handling as

well as the technique of abseiling from helicopters into jungle or mountain terrain. The intake for each Program is small, between 10 and 30 men, so it is possible to give instruction almost on an individual basis which results in a very high pass rate of 70 per cent on average. Those who pass are entitled to wear the Ranger flash and black beret and serve a normal tour of 24 months including four two-week leave periods, although the tour can be increased by up to six months under special circumstances. Rangers have to be young, fit and 'on the bounce', and rarely stay in the line over the age of 22 although many members who wish to re-enlist find a welcome reception in one of the other army corps and, as with the SAS, this helps in spreading knowledge and skills around.

The 75th Ranger Regiment is one of the select units which fall under US Special Operations Command. This has gradually evolved since

Face blackened with camouflage cream, a US soldier holds his sub-machine-gun at the ready.

the need for centralized control of special forces was realized at the beginning of the 1980s and is a rather strange organization which since 1987 has overseen the SEALs, the Rangers and the Special Operations Forces Groups – popularly known as the 'Green Berets' – but not the USMC or airborne and airmobile divisions. These instead form components in the Rapid Deployment Force. US Special Operations Command (USSOC) liaises directly with the Central and Defense Intelligence Agencies and National Security Agency (CIA, DIA and NSA), with their own close links to Britain's MI5 and 6, GCHQ and the SAS. The organization includes a psychological warfare and a civilian affairs corps. It is, you might say, a global trouble-shooter.

USSOC also controls various air force special operations squadrons which operate gunships for the support of the ground forces, electronic

Special Operations Forces – the 'Green Berets'

The role of the US Special Operations Forces is nowhere better expressed than in the words of their founding officer. Colonel Aaron Bank was a wartime veteran of the Office of Strategic Services (OSS) with extensive experience of clandestine operations behind Japanese lines in Burma. Against opposition from both the army and the newly-formed CIA, both of which thought they knew better, Bank formed a new school at Fort Bragg in North Carolina to train volunteers 'to infiltrate by land, sea or air, deep into enemy-occupied territory and organize the resistance/guerrilla potential to conduct special forces' operations with emphasis on guerrilla warfare'.

At this point, with the Cold War at its chilliest and heavy fighting continuing in Korea, Bank was one of the few who saw very early that the postwar world was going to see a revolution in unconventional warfare conducted by nationalistic or politically motivated forces who would disappear into the civilian background. The assistance of the local population in identifying them and helping to fight them would be extremely important.

Bank had very clear ideas about the sort of men he needed in his new force, whose first component was the 10th Special Forces (SF) Group (Airborne). He wanted trained veterans with airborne, Ranger or wartime special forces' experience who had proved themselves reliable and willing to learn new tricks. Because of their envisaged behind-the-lines role, he also wanted linguists and got a good response from European and Asiatic expatriates who knew they could acquire American citizenship by enlisting in the armed forces. In this respect the SF became something of an equivalent to the French Foreign Legion and still retains something of a multinational flavour. This is reinforced by mixed training exercises with the special forces of other nations.

interception and counter-measures aircraft and helicopters for the infiltration and exfiltration of US troops on covert or overt missions. Their duties also include giving exact navigational fixes for troops on the ground and maintaining communications in 'dead' radio areas. The USAF's main component is the 23rd Air Force, Military Airlift Command, based at Scott AFB in Illinois but with detachments all around the world. (The USAF has its own small elite ground force in the form of the 1st and 7th Special Operations Squadrons, the former based in the Philippines and the latter in West Germany. Its specific task is defence of American airfields from attack by Russian Spetsnaz troops and although virtually no details are available it can be assumed that its men are thoroughly versed in similar skills to those of the SAS.)

A SEAL team prepares for an airlift in a Lockheed C-141.

How the sobriquet 'Green Berets' was arrived at is a story in itself. General Bank was a staunch admirer of the British Commandos and thought that nothing could be more appropriate than for his new force to wear a similar piece of head-gear. This presumption of 'elitism' brought howls of rage from many senior army officers and its wearing was discontinued until 1961 when President Kennedy visited Fort Bragg and ordered its reinstatement! In mutual respect, after the President's assassination the Green Berets renamed their training school the John F Kennedy Special Warfare Center.

By this time there were three SF Groups, the 1st, 7th and 10th, and as American involvement grew in Vietnam there was further expansion until by 1964 there were seven, the newcomers being the 3rd, 5th, 6th and 8th. Each was given a specific geographical region as its particular concern. The 1st and 5th drew South-East Asia, including the Philippines and Taiwan, the 3rd Africa, the 6th the Middle East and the 8th Central and South America; the 7th saw varied duty in Germany, Vietnam and Central America while the 10th was divided between the continental United States and Germany. The 3rd saw service in support of government forces in the Congo, Ethiopia, Guinea and Kenya and the 6th in Iran, Jordan, Saudi Arabia, Turkey and Pakistan, largely acting as advisors to train indigenent troops in counter-insurgency tactics, but it was of course in Vietnam that the Green Berets came into the limelight. (They have more recently, of course, been involved in Colombia and Panama.)

The first men of the 1st Special Forces Group arrived as advisors in South Vietnam in 1957, shortly after the French withdrawal and the country's partitioning. Over the next five years the political situation gradually worsened and by 1962 there were some 4,000 American servicemen and women in Vietnam. At this point the Green Berets got involved in what was known as the Civil Irregular Defense Group programme

A US Ranger involved in jungle training.

designed to train the montagnards from the highlands in which the borders of Laos, Cambodia (Kampuchea) and South Vietnam met. These tribesmen do not class themselves as Vietnamese and were willing subjects for the Green Berets' 'hearts and minds' campaign, readily establishing fortified centres and eagerly adapting to modern smallarms despite an almost total lack of education. By 1964 some 18,000 of them were acting in a local defence and reconnaissance role for the US and South Vietnamese armies, operating out of fortified villages whose defences were progressively strengthened as the conflict escalated.

In this ame year the Green Berets introduced a new type of unit, the Studies and Observation Groups, a polite and innocuous name for units whose task was infiltrating the territory of surrounding countries on what were essentially espionage missions to pinpoint hostile troop build-ups. The 'SOGs' included SEALs and USMC Recon personnel as well as Green Berets. This development, coupled with Vietcong attacks on the mnontagnard villages, led in 1965 to the creation of Mobile Strike Force teams to attack Vietcong training camps over the borders. In both forms of mission, as well as in training 'loyal' Vietnamese, Laotian and Cambodian troops, the Green Berets acored a high success rate. They also undertook many reconnaissance, sabotage and ambush sorties as well as rescue missions – Son Tay on 21 November 1970 being the most publicized such attempt even though it was unsuccessful because of PoWs had been moved. When the last US Special Forces were finally withdrawn from Vietnam in 1971 they had earned 11,790 medals including 17 Medals of Honor, the US equivalent of the Victoria Cross.

From 1969 onwards there was a gradual rundown of Green Beret units. In that year the 3rd Group was deactivated, followed by the 6th in 1971, the 8th in 1972 and the 1st in 1974. With subsequent changes there are today four active

and four reserve SF units. The active ones are the reconstituted 1st based at Fort Lewis, Washington; the 5th at Fort Bragg; the 7th at Fort Benning and the 10th at Fort Devens, Massachusetts. One battalion from the 1st Group is stationed on Okinawa, one from the 5th at Fort Campbell, Kentucky, one from the 7th in Panama and one from the 10th at Bad Tölz, Germany. The reserve Groups are the 11th at Fort Meade, Maryland, the 12th at Arlington Heights, Illinois, the 19th (National Guard) at Salt Lake City, Utah,

ABOVE AND BELOW
Rangers wearing
protective facemasks on
the firing range with
M16s at Bad Tölz in West
Germany.

and the 20th (National Guard) at Birmingham, Alabama. All fall under the control of the US 1st Special Operations Command.

Each Group comprises three battalions which contains a headquarters company known as the 'C' Team and three companies known as 'B' Teams. Each 'B' Team is made up of five or six 'A' Teams, the Green Berets' normal tactical unit. These are larger than their equivalents in the SAS, being composed of 10 men plus a Lieutenant or Warrant Officer as executive officer and a Captain as CO. All the men are NCOs and each has his own speciality. There is an operations and an assistant operations sergeant, a heavy weapons and a light weapons leader, two medics, two radio operators and one engineer. As each member of the Green Berets has already had airborne or Ranger training, a description will not be repeated. Specialist training for the above roles lasts between 16 and 25 weeks although it is longer for medics – 43 weks including a month working in the casualty department of a hospital. Many SF personnel voluntarily take extra training in such skills as scuba diving, mountaineering, jungle warfare tactics, intelligence analysis and interpretation, demolition and escape and evasion. In fact, as in all special forces, training never really stops.

1st Special Forces Operational Detachment Delta

The rundown of SF forces in the post-Vietnam era worried many people in the Pentagon and elsewhere, especially with the very sivible increase in international terrorist activity. One Green Beret officer who had completed two tours in Vietnam after earlier service with the 82nd Airborne Division, and also served as an exchange officer with the SAS in 1962–63, shared these anxieties. In 1974 Colonel Charles Beckwith was appointed commandant of the John F Kennedy Special Warfare Center at Fort Bragg which out him in an ideal position to pro-

selytize his vision of an SAS-style counter-terrorist unit, and on 19th November 1977 he saw his wish fulfilled with the creation of 'Delta Force', as it is popularly known.

By 1979 enough volunteers from the Green Berets, 82nd Airborne and Rangers had passed the demanding induction course – which, based closely on the SAS model, surprised many recruits who already thought themselves perfectly fit and skilled – to fill two squadrons of approximately 100 men. Through security necessities, Delta's exact composition and strength are closely guarded secrets, but it is known that each squadron is divided into 16-man troops and further sub-divided into SAS-pattern four-man 'chalks'.

Unfortunately, Delta Force's first operational mission was a disaster, though not through any fault in the men or their training. In November

Men of the US 82nd Airborne Division during a training jump at Fort Bragg.

1979 militant Iranian supporters of the Ayatollah Khomeini stormed the American Embassy in Tehran and took 66 hostages. President Jimmy Carter ordered Delta Force to come up with a plan for their rescue, which was given the code-name 'Eagle Claw'. This took time because of the delicate political situation, the distances involved and the type of transportation to be used. In the end both the USMC and the Rangers became involved, the first to provide the necessary helicopters and the latter to establish safe perimeters around the two chosen landing zones. A combination of circumstances prevented the operation's success. The Rangers stopped a local bus and a petrol tanker on the road adjacent to 'Desert One' but a third vehicle escaped and the men of Delta Force knew they had lost the advantage of surprise. Then two of the Marines' Sea Stallion helicopters broke down,

meaning there was insufficient lift capability to extract the troops and the hostages. There was no option but withdrawal.

Since then the Delta Force has had one success, one tragic failure and two 'no-gos'. The first was the triumphant rescue of 79 passengers from a hijacked Venezuelan DC-9 in 1984 when Delta operated alongside Venezuelan special forces and killed two terrorists. On the second and third occasions, one involving a Kuwaiti airliner later in the same year and the other a TWA Boeing 727 in Beirut in 1985, Delta was alerted but not called in. On the fourth occasion in November that year Delta was called out when terrorists hijacked an Egyptian Boeing 737 but the Egyptian army jumped the gun, stormed the plane in Malta and caused 50 deaths amongst the hostages. Despite this rather gloomy record,

ABOVE Airborne infantry with an M60 7.62 mm machine-gun.

BELOW Part of the 505th Regiment, 82nd Airborne, at Sharma el Sheik in Egypt during a Multinational Force exercise in 1982.

none of which can be attributed to lack of professional skill in Delta's squadrons, the US government under both Presidents Reagan and Bush has continued to support the concept and will no doubt keep it operational.

The 82nd Airborne Division

Following the end of the Korean War all the US airborne divisions except the 'All Americans' were deactivated, although the 101st was reinstated as a helicopter-borne airmobile division in Vietnam. The 82nd next saw action on the island of Santo Domingo in 1965 when a revolution threatened the lives of American civilians, then in 1968 its 3rd Brigade was flown to Vietnam when the NVA launched their Tet Offensive which threatened to overrun the south. The brigade remained in Vietnam until 1969 when it returned to its Fort Bragg headquarters. More recently, two brigades were deployed in Grenada in 1983 where they operated alongside the Rangers to restore order and supervise a return to democracy.

The 82nd Airborne forms the spearhead of America's Rapid Deployment Force with one battalion permanently on 18 hours' standby to fly anywhere in the world with the rest of the division's Ready Brigade able to follow within a day. The division's three brigades rotate in the Ready role.

Each 4,000 men strong brigade contains three parachute battalions, but additional supporting units bring the division's total strength to some 18,000. The 1st Brigade contains two battalions of the 504th Infantry Regiment and one battalion of the 508th, the 2nd three battalions of the 325th and the 3rd two battalions of the 505th and one battalion of the 508th. There are three artillery battalions, the 319th, 320th and 321st, each with 18 105 mm guns and Gama Goat tractors, both of which can be air-dropped. Then there is the 3rd Battalion, 73rd Armored Regiment, with 54 M551 Sheridan light tanks. Although these vehicles are no longer used by

the rest of the army, they are retained by the 82nd because they can be parachuted in on pallets to give a useful anti-tank capability with their 152 mm Shillelagh missile launchers.

Other integral units in the 82nd are as follows: 82nd Signal and Combat Aviation Battalions, the latter with 98 Black Hawk troop-carrying helicopters, Kiowa reconnaissance and Cobra anti-tank helicopters, 82nd Military Police, Adjutant General and Finance Companies, 1st Squadron, 17th Cavalry, with a further 82 Black Hawks, Kiowas and Cobras, 14th Chemical

The 82nd Airborne Division land on drop zone Normandy in a training jump.

Detachment, 313th Military Intelligence Battalion, 3rd Battalion, 4th Air Defense Artillery with Vulcan cannon, 307th Medical Battalion, 407th Supply and Service Battalion and 782nd Maintenance Battalion, plus administrative units. From this it can clearly be seen that the division is a truly balanced all-arms force which in time of a confrontation with the Warsaw Pact could be in Germany within days. All of its equipment, including the helicopters, can be airlifted in Air Force C-141 Galaxy and C-5 Starlifter aircraft. For supply once the 82nd is on the ground C-130 Hercules can each drop up to 16 20,000 lb (9,000 kg) stores containers with ammunition, food and medical equipment. These employ 'smart' micro computer technology to steer themselves to exactly the right point, avoiding the hazard suffered by earlier parachute forces of having their supplies drop in the enemy lines.

ABOVE UH-60A Black Hawk helicopters spearhead a desert exercise by the 82nd Airborne.
OPPOSITE Men of the Canadian Special Service Force parachute into arctic terrain which is their speciality.

All military parachute training in America is conducted at Fort Benning in Georgia, the course lasting three weeks after, of course, each soldier has completed basic training. During Ground Week the volunteers learn the theory and practice of jumping, how to put on and release the harness and how to fall correctly. During Tower Week they move on to static line jumps from towers of progressively greater height culminating in 250 ft (76 m). Finally comes Jump Week when each recruit has to complete five jumps, one of them at night, from varying heights and using both the old T-10 and the army's current MC1-1B parachute. Over 21,000 men (and women) take the course each year and the instructors have a 95 per cent success rate. Not all go into the 82nd, of course, but all can consider they have joined America's elite forces.

CANADA

Special Service Force

The Canadian Special Service Force is an elite brigade-size formation dedicated to rapid deployment in support of NATO, particularly on the northern flank in Norway. The rugged, snow-covered terrain of northern Canada gives ample scope for mountain and arctic warfare training, and this is the brigade's speciality although it can fight anywhere in the world.

Its origins go back to the Canadian parachute battalion which formed part of the British 6th Airborne Division during the Second World War, and to the combined US/Canadian Special Service Force. Both were disbanded after 1945 apart from cadres which still practised parachuting and it was not until 1968 that the Canadian Airborne Regiment was created, and not until four years later still that the Special Service Force came into existence. With headquarters at St Hubert in Quebec, this includes an armoured regiment (8th Hussars) and an airmobile infantry battalion (1st Royal Canadian Regiment) plus the 2nd Royal Canadian Horse Artillery Regiment, engineer and signals units, but the brigade's principal 'teeth' are provided by the Airborne Regiment.

The regiment, 741 strong, is divided into a headquarters and signals squadron and three airborne commandos, the 1st English-speaking, the 2nd French and the 3rd bilingual. They are equipped with the usual gamut of M16A2 rifles, Belgian 7.62 mm and American .50 calibre machine guns and TOW anti-tank missiles. Training is very similar to that of the Royal Marines' Mountain & Arctic Warfare Cadre and the two units frequently conduct joint exercises with the Norwegians and Dutch, although the regiment is fully trained in both desert and jungle warfare as well. The regiment is on permanent 48-hour standby but has yet to see action. In a crisis situation, the rest of the Special Service Force could follow within 72 hours.

RIGHT The despatcher checks that each man's parachute static line is securely hooked on prior to a jump by the Special Service Force. **BELOW** Preparing for a night drop.

FEDERAL REPUBLIC OF GERMANY

1st Fallschirmjäger Division

When West Germany was admitted to NATO on 8 May, 1955 it was inevitable that the reconstituted army would include a Fallschirmjäger, or paratroop, element. The result was today's 1st Fallschirmjäger Division with headquarters at Bruschal. 'De-Nazified' officers and men from the wartime parachute regiment formed a training cadre and began looking for volunteers for the conscripts then being levied. Despite the strong anti-militarist tendencies amongst the youth of the time who had witnessed the horrors of total war as children, they were pleased at the response even though extra pay was offered as

Bearded West German paratroops with a Milan anti-tank missile.

an incentive. The modern division consists of three brigades, one being attached to each of the Bundeswehr's three Korps as a rapid deployment reserve. Each brigade consists of a headquarters and signals battalion and three airborne battalions. Within each battalion there are two rifle and two anti-tank companies, the latter equipped with TOW and Milan missiles.

Basic training takes three months after which the recruits – by this time in peak physical condition and well versed in fieldcraft and marksmanship – proceed to the parachute training centre at Altenstadt for the four-week course which culminates in five jumps, one at night. After this the men are entitled to wear paratrooper wings. Only a few elect to leave the army after their compulsory conscription period

of 15 months, most staying for a regular tour of 21 and many 'going career' and staying on for 15 years or even longer in the case of officers and NCOs. Most of the career soldiers progress after their first 12 months with a battalion to the advanced parachute school at Schöngau where they learn HALO and HAHO techniques, or to the Ranger school (also at Schöngau) to learn mountaineering and pathfinding. They practise urban warfare techniques, operating in teams of five men assaulting derelict factories and other buildings. Like the rest of the modern West German army, the paras are highly trained and skilled as well as superbly equipped, and consider themselves an elite.

1st Gebirgsjäger Division

The Gebirgsjäger, or mountain troops, Division, is the second truly elite formation within the total 12 divisions of the Bundeswehr. Constructed on classical lines with one armoured and two motorized infantry brigades, it also includes the 23rd Gebirgsjäger Brigade as its own special force specifically trained in guerrilla and counter-guerrilla warfare. They have a long tradition and amongst other campaigns, Gebirgsjäger fought alongside their Fallschirmjäger comrades during the battle for Crete in 1941. Three battalions strong, the 23rd Brigade is specifically entrusted with the defence of the Bavarian mountains and has a unique 'stay

Fun and games for German mountain troops as they struggle to load Milan components on to a long-suffering mule.

behind' role in that it is expected to melt into the mountains behind a successful Warsaw Pact thrust and seriously harass the invaders behind their own lines. For this reason, almost all volunteers for the brigade are locals long practised in skiing and mountaineering as well as surviving the hazards of blizzards and avalanches. 3,800 men strong (5,100 with reserves), the brigade is superbly trained in every aspect of mountain and arctic warfare. They are also the only troops in the modern West German army allowed to drink beer at lunchtime! Don't ask me why, they don't know themselves . . .

The Bundeswehr finally has three small squadrons of behind-the-lines reconnaissance troops drawn principally from the airborne and mountain brigades, the Fernspähkompanien or Long Range Reconnaissance Patrols, who are trained by the British SAS at Weingarten. One

Volunteers training for admission to GSG9; only one in five will make it.

squadron is attached to each Korps, their task in time of war being to operate in six-man teams in East Germany behind the Warsaw Pact line of advance. No other information is available, but they are organized and equipped on the model of the Danish Jägerkorps.

GSG9

The Federal Republic's crack anti-terrorist unit is not part of the army but is recruited exclusively from the police and Bundesgrenzschutz, or Federal Border Guards. Grenzschutzgruppe 9, generally just known as GSG9, was formed by Ulrich Wegener in the wake of the 1972 Olympic Games massacre of Israeli athletes by Palestinian terrorists. Although the West German government was reluctant to see the creation of such an undercover force, memories of the SS and Gestapo in mind, Wegener proved persuasive and soon had assembled a team of 60 volunteers. Working closely with the SAS and elite Israeli paras, Wegener devised a stringent 13-week training course which only 20 per cent of applicants succeed in passing. Recruits, already trained in German law, also learn the rudiments of international law in addition to marksmanship and all the usual special forces' techniques, including unarmed combat, the use of sniper rifles and silenced Heckler & Koch MP5s, communications and the use of eavesdropping equipment including directional microphones and cameras. Many also take the army's basic and advanced parachute courses. GSG9's strength currently averages between 160 and 200 men at any one time.

Vindication of the need for a unit like GSG9 came five years later when, in October 1977, four terrorists hijacked a Lufthansa Boeing 737 and demanded the release of 11 members of the notorious Baader-Meinhof gang from Stammheim prison. After the plane took off from Palma, Majorca, the hijackers broke into the flight deck and redirected it across the Mediterranean, first to Rome and then Cyprus to refuel, then on to South Yemen. Here, the landing was rather bumpy and the pilot, Captain Jürgen Schumann, asked permission to inspect the undercarriage for damage. Once outside the aircraft he bolted for the control tower to give Yemeni police descriptions of the hijackers and their weapons. The hijackers said they would blow up the aircraft if Schumann did not return immediately, which he did. Moments after take-off, the terrorist leader shot him in cold blood. When the aircraft landed at its next destination, Mogadishu in Somalia, the pilot's body was thrown out on to the runway. West German Chancellor Helmut Schmidt then gave the order to send in the GSG9 team which had already flown to Crete to await developments. With them were two advisors from the SAS.

With the cooperation of the Somali government and air traffic controllers, the GSG9 Boe-

GSG9 trains using an enormous variety of weapons and transport vehicles to make sure it is ready to tackle any hostage or hijack situation anywhere, at any time.

ing 707 touched down in darkness, all its lights extinguished. The commandos, dressed in black with soft rubber-soled boots, moved silently up to the hijacked 737, positioning themselves beside the emergency exits with aluminium ladders poised. At the pre-arranged moment, while other aircraft turned on their lights and revved their engines to distract the hijackers, the doors were blown out, the ladders slapped in place and the commandos stormed into the aircraft, yelling at the passengers to keep their heads down. It was over in seconds. Three of the hijackers were shot dead and one captured. The only other casualty was an air hostess wounded in the leg. It was a triumph both for Wegener's men and the West German government's stand against terrorism.

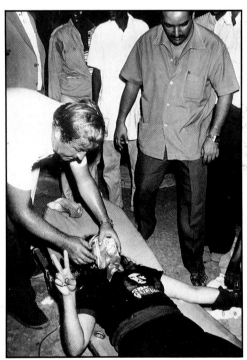

LEFT GSG9 was established to counteract the increasing wave of terrorism such as the bombing of a Pan Am jet at Rome's Leonardo da Vinci Airport in 1973 in which 31 people died.

ABOVE Still defiant although wounded and with her associates dead, female terrorist Suhaila Sayeh is treated by a doctor at Mogadishu after GSG9's assault.

THE NETHERLANDS

Marine Corps

Dutch special forces had a similar success when in May 1977 South Moluccan extremists hijacked the Rotterdam-Groningen passenger train at De Punt, demanding the creation of an independent state of South Molucca free from Indonesian rule. This was not the first such attempt for another train had been hijacked two years earlier, but on that occasion the terrorists had surrendered peacefully after a 13-day siege. This time, though, the extremists shot the train driver to show that they meant business. Despite this, the Dutch government played for time, stretching negotiations over nearly three weeks, giving the Marines of Holland's elite BBE (Bizondere Bystand Eenheid or 'Different Circumstances Unit') ample time to study the routines of the 13 hijackers and to study their faces through photos taken with telescopic lenses. The actual assault went in at dawn on 11 June, the terrorists being thrown off stride by a low-level pass by a pair of supersonic F-104 Starfighters whose noise and vibration stunned even those who were expecting it. Then the Marines blew in the train doors, screaming at the hostages to keep down. Anyone standing was an enemy and six of the Moluccans died in the shoot-out, the remainder being captured. Only one Marine was wounded but unfortunately two of the 80 civilian hostages were killed in the exchange of fire.

The BBE is part of the Marine Corps' 1st Amphibious Combat Group and is a select force of just 90 men organized in teams of five in three platoons, two active and one training. Volunteers, who must have already gone through the Corps' rigorous training course – virtually identical to that of the British Marines alongside whom they regularly exercise under command of 3 Commando Brigade – are closely studied to determine their stability and reactions under stress. Those deemed potentially suitable are

Rifle barrels protrude menacingly from the windows of a school where South Moluccan extremists held 110 teachers and pupils hostage for three days (top) at the same time as the train hijack (right), only releasing them when a virus swept through the children.

put through an intensive 48-week course with the third platoon during which, in addition to their existing skills, they learn counter-terrorist techniques in detail, including tactics for storming buildings, aircraft, ships and trains with minimal risk to hostages. They are also trained in riot control so they can be called in to assist the police in a crisis.

The principal fighting units of the Royal Netherlands Marine Corps, whose headquarters are in Rotterdam, are the two Amphibious Combat Groups, the 1st home-based with a principal dedication to the northern flank of NATO in Norway, the 2nd on the Antilles in the Caribbean. With supporting and independent units, total strength of the Corps averages 2,800 men. There are three independent units, 'Whiskey' Company, 150-strong, which forms a virtually integral part of 45 Commando, RM; the Company Boat Group, which is similarly integrated with 539 Assault Squadron, RM, and which specializes in amphibious operations in truly arctic conditions in the far north; and the Dutch SBS which is trained alongside the British SBS and includes advanced parachuting techniques, scuba diving and deep penetration reconnaissance missions in its repertoire. Despite their frequently shaggy haircuts and beards, the Dutch Marines are an elite force which has earned global respect.

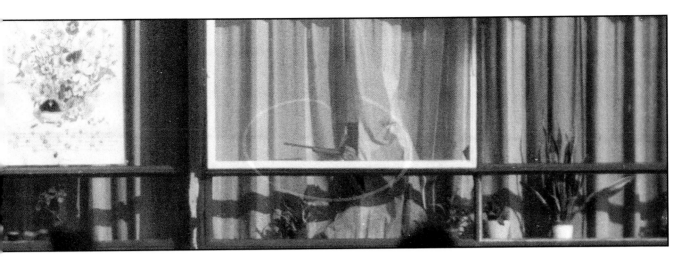

NORWAY

Jägers

Norway occupies a unique place in the NATO structure, its importance being determined by the commanding position it occupies to deter Warsaw Pact submarines and surface vessels from passing the North Cape or out of the Baltic into the Atlantic. This makes the country a prime target for the beginning of any Soviet offensive against the West – however unlikely that currently seems – and is the reason why so many elite mountain and arctic warfare units from other countries are dedicated to its defence. The whole Norwegian army is, of course, trained in M&AW, most of its population having learned to ski and climb from childhood. It is also unusual in that its vast bulk consists of reserves with civilian occupations, for national service only lasts a year. There is a permanent cadre of career personnel numbering about 13,500, swollen to 29,000 by each year's intake of conscripts, but on full mobilization Norway could field an army of a quarter of a million because each reservist keeps his uniform and weapons at home – a lesson learned from the German occupation during the Second World War. It would be unfair to say that the whole Norwegian army is an elite force, though in terms of specialist training and skills it is not far from the truth.

Given this, it is unsurprising that Norway only has two small units of special forces, both about 100-strong. These are the Marine and Para Jäger platoons – the word literally means 'hunter' – both of which share the same training as the Royal Marines' Mountain & Arctic Warfare Cadre, the former having similar specialties as the SBS and the latter the SAS. Inevitably, their principal roles are deep penetration reconnaissance and sabotage missions using canoes, sub-aqua gear or parachutes. There is, in addition, a 30-man police counter-terrorist squad modelled along GSG9 lines.

TOP A Leopard 1 armoured engineer and recovery vehicle in use with the Norwegian army.

ABOVE Men of the crack Norwegian armoured battalion 'Nord' with a Leopard 1 main battle tank.

OPPOSITE The Norwegian army helps train British and other NATO special forces in mountain and arctic warfare.

DENMARK

Fromankorpset

Denmark only has a tiny standing army, smaller in size than an American division, organized in five brigades, but each of these has an attached parachute commando unit of 10 men: the Fromankorpset or, as they are more generally known, the Jägerkorps. Established in 1961, this small body of men parallels the British SBS, and all, of course, are volunteers who undergo a rigorous two-year training programme in deep reconnaissance, small boat handling, scuba diving and parachuting. They are also trained in counter-terrorist tractics and would be used if necessary against hijackers.

BELGIUM

Le Régiment Para-Commando

With origins going back to the wartime SAS Brigade and Commandos, the Belgian Para-Commando Regiment has an illustrious history augmented since its official formation in 1952 by a variety of actions, notably in Katanga (now Shaba) province of the Congo (now Zaire) in 1960 and 1964. In particular, in the latter year they parachuted into Stanleyville airport (now Kisangani) to secure it against rebel Simba forces and provide time for the evacuation of the White Belgian population; they were later also involved in the rescue of civilians from Kolwezi, also in Zaire, alongside French Foreign Legion paras.

The regiment, consisting of three battalions plus an anti-tank company currently equipped with Milan missiles, a battery of 105 mm guns and a mechanized reconnaissance squadron, is trained in both amphibious and parachute assault and maintains close links with the SAS and SBS with which it shares similar recruiting and training standards. Within NATO, the regiment's principal role is as part of the AMF (Allied Command Europe Mobile Force), but it is dedicated primarily to the southern (Balkan) flank

OPPOSITE General Charles de Gaulle decorates men of the French Foreign Legion in North Africa during World War II. The Legion still provides the backbone of the French rapid deployment force.

instead of to Norway. Belgium also has a dedicated counter-terrorist unit, the Escadron Special D'Intervention, but little is known of this.

FRANCE

Force d'Action Rapide

Like the United States, France has in recent years created its own rapid deployment force to act both as part of NATO in case of a confrontation with the Warsaw Pact, and to safeguard French interests elsewhere around the world. It consists of five crack army divisions, some 47,000 men in all, plus a number of smaller units including the French Foreign Legion's famous 2e Régiment Etranger de Parachutistes (2e REP – 2nd Foreign Parachute Regiment).

The nucleus of the force is the 11e Division Parachutiste which was created in 1971, the earlier 10th Division being disbanded following the abortive coup in 1961 which failed to prevent Algeria being granted independence even after the paras had played a key role in restoring General Charles de Gaulle to power. French paras have seen a great deal of action since the first unit was formed in April 1948, particularly in Indo-China where they suffered horrific casualties during the battle of Dien Bien Phu against General Giap's Viet Minh. After Algeria, when the Foreign Legion's 1st Parachute Regiment was also disbanded for its part in the revolt, France's sole surviving parachute unit for 10 years was 2e REP, which saw action in Chad and Djibouti.

The current 11e Division Parachutiste averages 13–14,000 men, divided into six regular army regiments plus the 2e REP, a light mechanized regiment, a regiment of artillery, one of engineers and headquarters and logistics units. In addition there is the 1èr Régiment Parachutiste d'Infanterie de Marine, a naval parachute commando unit trained to the same sort of standards as the SBS or SEALs and tasked with similar duties. Both this unit and 2e REP are fully trained parachutists and skilled scuba divers,

mountain and arctic warfare experts, snipers and communications specialists well capable of completing deep penetration reconnaissance and sabotage missions. They are also France's principal counter-terrorist force.

The division showed its rapid deployment capability in 1978 when Congolese rebels seized the mining town of Kolwezi, in Zaire. The 1,500-strong FNLC force immediately commenced an orgy of looting and rape and set up summary courts to deal with anyone suspected of being a mercenary. Several people were executed but the Belgian government was reluctant to employ direct action itself to protect the 2,300 Belgian, French, Italian and Portuguese employees of the Union Minière and their families. They appealed to the French government and on 17 May 2e REP was alerted. There was controlled pandemonium at the regiment's headquarters, Camp Raffali outside Calvi on the island of Corsica, as Lieutenant-Colonel Erulin assembled his forces.

LEFT The French Force d'Action Rapide uses a wide variety of armoured and unarmoured vehicles to give it necessary mobility.

ABOVE French paras load bazookas on to a Jeep during the rescue operation at Kolwezi.

French Foreign Légionnaires embark on a landing craft for an exercise. One of their most unusual weapons, particularly in jungle fighting, is the crossbow.

The first four companies flew out to Kinsasa on the morning of the 18th, the remainder of the regiment with the vehicles and heavy weapons following in US Air Force transports. To save space in the first wave of aircraft, the men had left their parachutes behind, and when they were issued with American T-10 rigs they found the harness incompatible with their jump bags, so had to improvise leg straps to hold them and tie their weapons on with cord wherever they could. Hardly an auspicious start! Nevertheless, by mid-morning on the 19th three companies of paras managed to cram themselves into the four Hercules and one C-160 aircraft available and lifted off for Kolwezi. The fourth company would follow later.

After a hot, exhausting four-hour flight they arrived over the town and began dropping from the sky. They rapidly seized key positions against sporadic opposition but had to beat off a fierce counter-attack during the night. The fourth company arrived in the morning of the 20th, together with the regiment's mortars, and the reinforced paras made short work of the rebels, chasing them into the surrounding countryside after killing 250 of them, and released the imprisoned Europeans. Their own casualties numbered just five killed. The légionnaires stayed on for a few days to help restore order but were back in Corsica by 4 June.

Today the regiment comprises about 1,300 officers and men divided into six companies: command and services, reconnaissance and support and four rifle companies. As in the SAS 'Sabre' squadrons, each of these four companies has a specific role. One is anti-tank but

also specializes in urban warfare; the second is mountain and arctic, the third amphibious while the fourth specializes in sniping, demolition and sabotage. This pattern is repeated throughout the six regular regiments of the 11e Division Parachutiste.

The other components of the Force d'Action Rapide are as follows: 4e Division Aeromobile (4th Airmobile Division), 6e Division Légère Blindée (6th Light Armoured Division), 9e Division d'Infanterie Marine (9th Marine Infantry Division) and 27e Division Alpine (27th Alpine Division). The 4th Airmobile Division consists of 6,400 men equipped with a mixed force of 214 helicopters in the troop-carrying, anti-tank and reconnaissance roles. Between them they mount 400 anti-tank missile launchers and the infantry themselves have a further 48 Milan fir-

A FFL river patrol in Guyana. The men are armed with 5.56 mm FA MAS rifles.

Training is both hard and exhausting for men of the French rapid reaction force.

French paras deploy in mountainous terrain from an SA 330 Puma helicopter.

ing posts. The 6th Light Armoured Division contains 7,500 men divided into seven regiments, two equipped with 36 AMX10 armoured cars and two with armoured personnel carriers, 24 of which are fitted with HOT anti-tank missiles. The other three are the headquarters, artillery and engineer regiments.

The 9th Marine Infantry Division consists of 8,000 men similarly divided into seven regiments, all trained and equipped for amphibious operations. There are three regiments of mechanized infantry, one armoured, one artillery, one head-quarters and one engineer regiment. The Alpine Division is the strongest of all except the 11e Division Parachutiste, with 9,000 men in the standard seven regiments, all fully qualified in every aspect of mountain and arctic warfare. Six of the regiments are infantry, the seventh being equipped with armoured cars, and the division possesses no fewer than 108 Milan firing posts.

The Force d'Action Rapide is a tough, 'go any-where' corps whose anti-tank capability in particular makes it an invaluable asset to NATO.

ITALY

COMSUBIN

The Commando Raggruppamento Subacqui ed Incursori, or Sub-aqua Raider Commando Group, is Italy's equivalent of the SBS or SEALs and is recruited in much the same way from the Marines. The Italians have a long tradition in underwater warfare techniques and scored many successes against the Allies during the Second World War, with audacious attacks on shipping in the closely defended harbours of both Gibraltar and Alexandria. Indeed, it was the Italians who pioneered the whole concept and technology of 'frogmen' and 'human torpedoes'.

The men of today's COMSUBIN number about 200 and have a counter-terrorist role in addition to their primary military tasks of beach reconnaissance, mine clearing and sabotage of enemy port installations. All the unit's members have to go through a strenuous 10-month course of physical training, endurance tests and marksmanship, etc, followed by 42 weeks of Ranger and airborne training before qualifying. Their favoured weapons, as throughout the Italian special forces, are the Beretta 9 mm Model 12s sub-machine gun and AR70 assault rifle.

LEFT Amongst its elite forces Italy also counts its five alpine brigades, whose principal task is the defence of the mountainous borders with Austria and Yugoslavia although one brigade is regularly stationed in Norway.

OPPOSITE Ominously garbed members of COMSUBIN with a Beretta automatic pistol and Franchi SPAS M12 combat shotgun.

Beretta M12 and AR70

The name 'Beretta' has been synonymous with high quality smallarms since 1918 and during the Second World War both German and Allied troops went out of their way to gain pistols and sub-machine guns bearing their hallmark. The 9 mm M12 sub-machine gun first went into production in 1959, but has subsequently been improved, particularly by means of a new selector switch giving single-shot, three-round burst or fully automatic capability, and by coating the exterior with epoxy resin to prevent rusting. It is a very compact weapon with a folding stock, smaller than the Heckler & Koch MP5 or Israeli Uzi favoured by other elite forces, but the Italians like it because the design of the magazine receiver prevents mud or dirt clogging the mechanism. It can be fitted with either 20-, 32- or 40-round box magazines to give a cyclic rate of fire of 550 rounds per minute with an effective combat range of about a hundred yards.

The AR70 is one of Beretta's more recent designs and utilizes the new NATO standard 5.56 mm round. It is a well made, lightweight weapon with a 20- or 30-round magazine and is so accurate to about half a mile that a special telescopic sight can be used to turn it into a sniper rifle. It is normally fitted with a nylonite butt although special forces tend to prefer the folding tubular metal version. Its muzzle incorporates an adapter for firing 40 mm grenades. Even though neither of these weapons has so far been adopted by any other NATO army, the Italian elite forces are very happy with them.

OPPOSITE As well as being delpoyed by sea as here, the San Marco Marines are fully trained in parachute and helicopter assault techniques.

San Marco Marines Battalion

The 1,000-strong San Marco Marines Battalion, actually a regiment in all but name, is an amphibious task force dedicated to the protection of NATO's southern flank – *ie,* Greece and Turkey. Italy has had conscription since 1947 but all members of the battalion are volunteers, as you would expect. Their rigorous training covers all aspects of landing from their two ex-US Navy landing ships whether in boats or LVTP-7 Amtrack amphibious armoured personnel carriers, and as well as beach assaults their training includes cliff climbing. All members of the battalion are also parachutists and experienced in helicopter usage. The battalion is organized in three sections, operations, training and logistics, the operational section comprising four rifle companies. In recent years the battalion has formed part of the United Nations peacekeeping force in Beirut.

Despite their all round versatility, the San Marco Marines are specifically dedicated to the amphibious role and regularly exercise using landing craft, small assault boats, rubber inflatable dinghies and canoes.

Folgore Brigade, Nucleo Operativo Centrale di Sicurezza and Groupe Interventional Speciale

Formed in 1952 around a nucleus of wartime paratroopers and given its present title in 1978, the Folgore Brigade is Italy's principal airmobile force, being equipped with both fixed-wing aircraft for parachute drops and helicopters. As with the San Marco Marines, the brigade's principal role in time of war would be on the southern flank of NATO so in addition to normal infantry and parachuting skills, its men – all volunteers – are also highly trained in mountain warfare tactics and techniques, including skiing. Training lasts 16 months, three of which are devoted to learning the use of the 105 mm M56 Pack Howitzer. The Brigade consists of an engineer company, aviation flight, engineer flight, artillery battalion and a parachute infantry in north Africa, one in the Canary Islands and the fourth at home. It includes a small commando-style Special Operations Unit at Ronda which comprises selected volunteers who train in parachuting and scuba diving.

GREECE

Within Greece's conscript army there is a small parachute commando brigade trained in either parachute or helicopter assault and in mountain warfare, as well as an alpine raiding company. Counter-terrorist work is entrusted to a 50-man police unit based in Athens.

PORTUGAL

Portugal was late in developing a modern elite unit but began building up the current Special Forces Brigade in 1984. It is principally composed of veterans from the wars in Angola and Mozambique and consists of two light infantry battalions who are parachute-trained plus headquarters, signals, engineer and support companies.

The Warsaw Pact

Air Assault Force

Soviet airborne forces, the Vozdushno-Desantnaya Voyska or VDV, are the strongest in the world, consisting of more than 100,000 men in eight parachute divisions, eight air assault brigades, eight airmobile brigades and a variety of smaller units including Spetsnaz special forces. As seen in the first chapter, Russia was a pioneer in the use of airborne forces and even though they were not successful in this role during the Second World War, they were retained and in 1956 transferred from air force to army control. After the Cuban missile crisis, in 1964 they were shifted to the direct command of the Defence Ministry as what amounts to a separate branch of the armed forces and in time of war would be deployed by the Soviet high command STAVKA either as a rapid mobile force or as a strategic reserve, depending on circumstances. If the Soviet Union ever did invade Western Europe, the airborne forces would be used strategically to seize crossings over the river Rhine and elsewhere up to 300 miles in front of the advancing armoured divisions, and tactically in smaller force to capture airfields and knock out allied headquarters and communications centres. They would also be used to 'mop up' in the wake of a nuclear strike and to this end Soviet paratroopers are trained more intensively than those of any other nation in large scale exercises wearing protective clothing and masks against nuclear, chemical and biological contamination.

The Soviet armed forces are composed almost entirely of conscripts. Many from provinces such as the Baltic States, Azerbaijan, Georgia, Moldavia and the Ukraine are extremely reluctant conscripts – indeed, large numbers of their parents fought for the Germans during the Second World War. Within the airborne units the Russians have nevertheless succeeded in

Soviet paratroopers in full dress uniform parade through Red Square on May Day. They are carrying AKS-74 assault rifles with folding stocks.

building up a surprising *esprit de corps* even though their confidence was as shaken by experience in Afghanistan as American morale was affected in Vietnam a decade earlier.

Most conscripts in the Soviet armed forces will have already undergone rudimentary military training in DOSAAF, the schoolboy 'Voluntary Society for Cooperation with the Army, Air Force and Fleet'. This is not quite a cadet corps in the Western sense, more a 'militarized' Boy Scout movement. It encourages athletic ability and teaches basic smallarms usage as well as offering parachute and scuba diving courses, sailing, canoeing, skiing and rock climbing. All teenagers have to undergo a minimum of 140

By Western standards the quality of Soviet airborne equipment and training is poor, but sheer numbers make up for these lacks. ABOVE Obviously tense troops prepare for a static line jump while (opposite) a more experienced para exhilarates in the thrill of a free-fall dive.

hours' 'voluntary' training in these paramilitary skills with DOSAAF. A keen youngster will therefore be more fit and attuned to military requirements when he is called up at the age of 18 than his counterpart in any Western country. This facilitates training despite the enormous racial, religious and linguistic differences within the Soviet Union that have threatened to tear the central apparatus apart for decades and finally seem, at the time of writing, to be verging on success.

Members of the airborne forces are in any case all volunteers from within each annual conscript intake but have to endure the same four weeks' basic training designed as in every army to both toughen them up and teach them discipline. That discipline is fairly brutal and specifically intended to force independently-minded 'loners' into line, for the Soviet armed forces do not encourage either introspection or initiative at this early stage, only rigid obedience. This has long been a Soviet failing due to the precarious hold on loyalties which the Communist Party exerts, and is only slowly changing in line with other developments in Soviet society as a whole.

Given this, it is not surprising that some 85 per cent of volunteers for the airborne forces are members of Komsomol, the Communist youth movement, with already affirmed loyalty to the State. The remainder are encouraged by the same sense of adventure that leads so many to take up parachuting and earn the chance to be considered a member of an elite. It is not by chance that Soviet airborne forces have always been to the forefront of foreign deployments, as in Czechoslovakia in 1968 and Afghanistan from 1979 until recently, since they are generally regarded as being amongst the most politically reliable troops in the Soviet armed forces.

After their four weeks' 'basic', volunteers for the airborne forces are assigned to either the 44th Guards Airborne or 106th Guards Air Assault Training Division for specialized tuition in parachuting and helicopter air-landing techniques and tactics. By the standards of Western elite forces, this tuition is crude and rudimentary, but the Soviet Union has always relied upon its massive manpower to achieve results through quantity rather than quality.

Apart from the two training divisions, which would quickly become active in time of war, the other six are kept at a permanent state of readiness for rapid deployment anywhere in the world. These are the 7th Guards based in the Baltic states; the 76th Guards based at Pskov in the Leningrad Military District; the 98th Guards in Odessa; the 103rd Guards, which bore much of

the brunt of the fighting in Afghanistan, in Belo-russia; the 104th Guards in the Transcaucasus District; and the 105th which also fought in Afghanistan, in Turkestan. The 44th Guards are also based in the Baltic Military District and the 106th just outside Moscow; there is clear evidence that one role envisaged for both of these formations would be quelling any major civil disturbances.

Each division consists of 6,500–7,200 officers and men in three mechanized infantry regiments plus an artillery regiment, battalions of anti-tank, engineer, signals, maintenance, logistics and medical troops, a chemical defence company – actually a chemical warfare company although this is not admitted – and a parachute rigging company. The principal infantry weapons are the AKS-74 assault rifle and AKR sub-machine gun plus the RPK-74 squad support weapon.

Soviet paratroops accompanied by a BMD – a troop-carrying, air-portable and amphibious vehicle armed with both a 73 mm gun and an anti-tank missile launcher to give the airborne arm battlefield mobility and fire support.

A noteworthy feature of the Soviet airborne infantry regiments is that they are completely mechanized, each division having 330 BMD armoured personnel carriers. These fully-tracked, amphibious vehicles are capable of being dropped by parachute and are specifically designed to operate on the nuclear battlefield, being provided with full NBC filtration equipment. They carry a squad of six infantrymen plus the driver, commander and gunner, the latter sitting in the small turret which is fitted with both a 73mm smoothbore gun and a launcher for AT-5 anti-tank missiles. This gives the Soviet airborne forces much better battlefield mobility than those of most other nations as well as excellent anti-tank capability: an important asset to troops who may have to cope independently with enemy armour before their own tanks arrive. There is also a turretless

AK/AKS/RPK-74 and AKR

The VDV always receives the latest equipment before the rest of the army, and when Soviet designers belatedly began to copy the NATO trend towards smaller-calibre infantry weapons in the early 1970s, they were the first to be issued with the new AK-74 assault rifle. This is a greatly modified version of Mikhail Kalashnikov's famous AK-47 (both numbers refer to the date of entry into service) which has been produced in untold millions and is the favoured weapon worldwide of guerrilla and terrorist groups. The AK-47 and its immediate successor, the AKM which is of simplified and therefore cheaper construction, are chambered to 7.62 mm but the AK-74 fires a hollow-point 5.45 mm bullet. Although smaller, this will inflict a hideous wound and from the user's point of view is preferable because of the reduced recoil, especially in fully-automatic mode. It should be pointed out that international convention prohibits the use of hollow-point or 'dum-dum' bullets in military weapons although, strangely, they are permitted and indeed favoured for big game hunting because they give a 'cleaner' kill.

The AK-74 is very similar outwardly to the earlier AK-47 but is lighter and has both a higher rate of fire and greater muzzle velocity (650 rounds per minute and 2,955 feet per second compared with 600 and 2,330). The weapon is fed by a curved 30-round box magazine which slots into the receiver mechanism in front of the trigger. The basic AK-74 has a conventional stock but the AKS-74 favoured by the airborne forces has a folding tubular stock. The RPK-74 is a light machine gun version with a bipod support, a longer and heavier barrel and a 40-round magazine. The AKR sub-machine gun is virtually identical to the AKS-74 but with a barrel only half the length (7.87 instead of 15.75 inches or 200 instead of 400 mm) is especially suited for use by the crews of armoured fighting vehicles. Like all sub-machine guns, it only has an effective range of 100 yards or so though, compared with the rifle's 500 yards (550 m) plus.

A standard AK-47.

command variant with map tables and extra communications equipment.

Parachute forces are transported into action in Antonov An-12 turboprops – Russian equivalent of the Hercules – or the more recent and larger Ilyushin Il-76 turbojet. The former can each carry up to 90 troops (only 60 if wearing parachutes) or two BMDs, the latter 120 men or three BMDs. In addition there is a smaller number of the even larger An-22 turboprop which can carry 175 men or four BMDs. The air assault brigades, each of which comprises 2,000–2,600 men, are parachute-trained but would normally go into action in Mi1 Mi-6, -8 and -26 helicopters. Anti-tank support would be provided by Mi-24s with four missile launchers or the more recent Mi-28. The airmobile brigades are not parachute-trained. Weaker than the air assault forces with between 1,700 and 1,850 men, they would be flown in by conventional aircraft or helicopters after the paras had seized a suitable landing strip.

Naval Infantry

Although the 40 wartime naval infantry (Morskaya Pekhota) brigades were disbanded in 1947, when Admiral Sergei Gorshkov began rebuilding the Soviet Fleets in the late 1950s and early 1960s he realized the need for specially trained amphibious forces, especially for operations on NATO's northern and southern flanks in Norway and Turkey. Accordingly he authorized the formation of four new regiments of naval infantry, one to be attached to the Northern Fleet at Pechenga, one to the Baltic Fleet at Baltysk, one to the Black Sea Fleet at Sevastopol and one to the Pacific Fleet at Vladivostock. Subsequently, a fifth regiment has been raised, also based at Vladivostock. The Pacific garrison is now classified as a division while the other regiments have been increased to the size of approximately 3,000-strong brigades. Total strength of the naval infantry formations is estimated at some 16–18,000 men.

As in the airborne brigades, most are conscripts but they are selected from the fittest and most suitable recruits for they are also considered an elite and serve for three years instead of two as in the army. Basic training lasts nine weeks followed by four months of specialist instruction in such subjects as beach reconnaissance and cliff-climbing, communications, heavy weapons' usage, driving and first aid. Many men take further voluntary courses in parachuting, scuba diving and demolition work. Particular emphasis is placed on close combat and recruits receive intensive instruction in hand-to-hand fighting and the use of knives and bayonets.

Each brigade consists of three mechanized rifle battalions with amphibious BTR-60 armoured personnel carriers plus a tank battalion with a mixture of 34 PT-76 light amphibious tanks and 10 T-72 main battle tanks fitted with schnorkels for deep wading. The Pacific 'division' has five rifle and two armoured battalions. In addition there are specialized anti-tank, reconnaissance, rocket launcher and air defence

OPPOSITE BTR-60PBs, eight-wheeled armoured personnel carriers used by the Soviet naval infantry. **ABOVE** Naval infantry leave a landing ship in BTR-50 tracked armoured personnel carriers. **LEFT** Combat frogmen of a naval engineer detachment, Russian equivalent of the UDTs.

companies and mortar, medical, signals, supply and chemical 'defence' platoons. They are carried into action in a variety of purpose-built ships including two large 14,000-ton LPDs (Landing Platform Docks) each capable of carrying an entire battalion with all its equipment. There are also 32 smaller LSTs (Landing Ship Tanks) and 60 hovercraft of various types, the largest of which can carry four PT-76s or BTR-60s in a covered well deck. Although small in numbers compared to the US Marine Corps, the Soviet naval infantry are well-trained and skilled troops and more than a cut above the ordinary army infantry regiments.

Spetsnaz

The Voyska Specialnoye Naznachenia, generally abbreviated to Spetsnaz, is the Soviet Union's elite special force and roughly anala-

Naval infantry in a beach assault demonstration.

gous to the SAS and SBS since it includes personnel drawn from both the airborne and the naval infantry regiments. However, Spetsnaz is far larger than the special forces of any other country with at least 15,000 men at the most conservative estimate. This causes NATO officials such concern, particularly in a European context, that regular military exercises are held to practise tactics for dealing with them. In these exercises 'friendly' special forces such as the SAS are given the task of penetrating to assigned targets defended by regular army, navy and air force personnel supported by the police and civil defence authorities. Despite the fact that the defenders appreciate the point of the exercises and are fully on the alert, it is a worrying fact that the intruders generally manage to achieve their tasks.

Spetsnaz has essentially the same role as other special forces – reconnaissance and sabotage missions deep behind enemy lines – but it also has other roles, including intelligence gathering and assassination. There is mounting

evidence that Spetsnaz teams have already been active, not just in Western Europe and Scandinavia but also in Japan and the Philippines, for several years. The reason is that Spetsnaz is controlled by neither the army nor the navy, but by the GRU, the Glavnoe Razvedyvatelnoe Upravlenie or Main Intelligence Directorate of the General Staff. The GRU is Russia's military intelligence organization and co-exists somewhat uneasily alongside the KGB whose interests are more political, but its activities are far broader than this description suggests. The Soviet Union is starved for Western hi-tech know-how and one of the GRU's principal tasks is spying out industrial secrets, particularly in the electronic and computer fields. Spetsnaz is one of the tools the Directorate employs to secure this type of information.

Spetsnaz has wartime origins as an elite body of men (and women) trained like the British Special Operations Executive and American Office of Strategic Services to provide support in the form of communications, weapons training and sabotage techniques for resistance groups in German-occupied territory, and Russian partisans were particularly effective in tying down thousands of German troops in security duties. After the war an army officer, General Viktor Kharchenko, succeeded in persuading the high command, STAVKA, to retain these special purpose forces and turn them into an elite equipped with the latest and best weaponry and communications equipment.

Their role is described by the *Soviet Military Encyclopaedia* – a Russian reference book – as 'reconnaissance carried out to subvert the political, economic and military potential and morale of a probable or actual enemy. The primary missions of special reconnaissance are: acquiring intelligence on major economic and military installations and either destroying them or putting them out of action; organizing sabotage and acts of subversion; carrying out punitive operations against rebels; conducting propaganda; forming and training insurgent detachments, etc.' To these ends Spetsnaz forces not only spearheaded the military operation in Czechoslovakia in 1968 but also attacked President Hazifullah Amin's palace at the beginning of the invasion of Afghanistan in 1979, killing him and most of his guards and later operating in the mountains against the Mujahideen. They also operate as instructors for terrorists and in time of war would liaise with left-wing subversive groups already in place in Western countries.

There is firm evidence that Spetsnaz personnel are highly active in Europe, using the TIR convention to masquerade as the drivers of long-distance lorries with bonded cargoes which are frequently seen parked for prolonged periods outside NATO bases or trailing military convoys during exercises, particularly in West Germany. They also pose as merchant seamen, needing no passports to freely enter any NATO country with only the most cursory customs checks, which gives them access to naval

installations. They have been spotted wearing scuba gear and using miniature submarines scouting out Swedish and Norwegian coastal defences. And even more worryingly, they have been implicated in the mysterious deaths of a growing number of British scientists working in the defence and computer industries. The motive here seems to be to silence people who have refused to respond to blackmail or other incentives for espionage, or to hamper research efforts in critical fields by disposing of key personnel. A recent book (*Operation Spetsnaz*, published by Patrick Stephens Ltd), written by former SAS trooper Mike Welham and the author, presents the detailed case for these assertions.

Spetsnaz recruits, all Komsomol members with DOSAAF experience, are selected from the cream of each year's conscript intake, the GRU being given priority over even the strategic rocket, nuclear submarine or airborne forces. After three months' basic training, those who have proved themselves suitable are sent on a gruelling six-month NCO course. They do not just have to learn normal soldiering skills – smallarms usage, fieldcraft, enemy equipment recognition, first aid, NBC protection, etc, but a great deal more. Instruction in map reading, cross-country navigation and living off the land are obviously essential but Spetsnaz candidates also have to learn foreign languages (English

The Czech VZ61 Skorpion machine-pistol used by members of Spetsnaz.

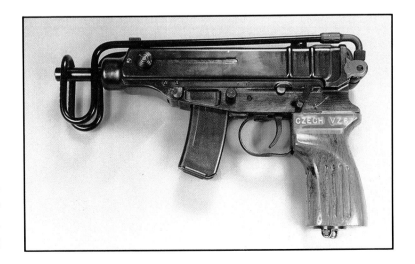

and German being the most common), the use of NATO smallarms, explosives and demolition techniques, unarmed combat, signals, codes and cyphers, interrogation techniques and how themselves to resist interrogation. Mountain and arctic warfare tactics and techniques and free-fall parachuting are also taught.

On completion of their training, graduates are assigned to a Spetsnaz unit for the duration of their service period. One Spetsnaz brigade is assigned to each of the army's 16 Fronts (a Soviet Front is analagous to a Western Army Group) and one to each of the four Fleets. The naval Spetsnaz brigades are composed of recruits who are additionally taught the tasks of scuba diving, beach reconnaissance and underwater demolition. In addition there is a Spetsnaz intelligence company to each Front and Fleet and three long-range reconnaissance regiments, one in the Far East and two in western Russia for deployment in Europe. Each brigade is divided into three or four battalions, each battalion consisting of three companies with 20 officers and warrant officers and 95 men. Additionally, there are headquarters and signals companies and an anti-VIP company whose 70–80 men are trained assassins tasked in time of emergency with killing Western political, military and industrial leaders.

Apart from deep penetration reconnaissance missions, Spetsnaz teams would be specifically tasked with eliminating British, French and American tactical battlefield nuclear missile carriers such as Lance and Pluton, with knocking out radar installations and communications centres, attacking airfields housing nuclear strike aircraft and with sabotaging industrial plants, radio and television stations and railway junctions. Spetsnaz personnel are the highly trained, motivated and skilled as well as the best equipped within the entire Warsaw Pact and many people feel they present a significant potential threat despite all the recent changes in the political climate.

GERMAN DEMOCRATIC REPUBLIC

With the wind of change blowing through the Soviet bloc countries, frontier barriers going down, free trade with the West and genuine elections for the first time in decades, it is difficult to see how the Warsaw Pact will hold together much longer when there is no further a need for it. What may emerge is a form of mutual defence pact no longer actively directed against the West but just there 'in case', for old suspicions die hard. Some or all of the nations covered in this section may, indeed, follow Austria's example and opt for neutrality. At the present, though, all that can be done is review the composition and status of the Warsaw Pact elite forces as they existed in 1989. What is certain is that, as in the West, the need for elite counter-terrorist units will remain for the foreseeable future.

The relatively small East German Nationale Volksarmee (National People's Army) only has a single airborne unit, the 40th 'Willi Sanger' Battalion which was formed in 1973. Precise details of its organization are unknown but it is believed to be trained Spetsnaz-style for behind-the-lines reconnaissance missions during which it would liaise closely with East German 'sleeper' agents in West Germany. The existence of a large number of subversive cells in the Federal Republic has long been acknowledged and however welcome it may be to the general public, the tearing down of the Berlin Wall poses a security nightmare to the West German intelligence services.

The Willi Sanger Battalion is trained alongside recruits in the HVA, East Germany's principal external intelligence agency, at the Edgar-Andrae Ausbildungszentrale (training school) which is hidden deep in a forest near the garrison town of Lehnin, and the battalion has its headquarters at Proro on the island of Rugen in the Baltic. Also based at Proro is the 29th 'Ernst

Moritz Arndt' Regiment, a mechanized infantry unit trained like the Soviet naval infantry in amphibious warfare. Finally, the navy includes a small number of combat swimmer/diver teams for beach reconnaissance and underwater demolition work.

POLAND

Largest of the Warsaw Pact armies other than that of the Soviet Union itself, the Polish People's Army (Ludowe Wojsko Polskie) contains one airborne and one amphibious assault marine division, each of which includes a small special forces component. Formed in 1957, the 6th Pomeranian Air Assault Division is a parachute-trained but increasingly heliborne 'division' of reinforced brigade size with about 4,000 men in five battalions, one of which is mechanized and one dedicated to Spetsnaz-type behind-the-lines reconnaissance and sabotage operations, probably in Holland. At least one of the brigade's battalions took part in the invasion of Czechoslovakia in 1968. As in the Soviet Union, most of the division's members are already in the Communist Party when conscripted and will have learnt parachuting in youth clubs. The brigade is known to practice mountain and arctic warfare in the Carpathians.

The 7th Luzycka Naval Assault Division based at Gdansk is really just a reinforced brigade with some 5,500 men. An army rather than a navy unit, it consists of three rifle regiments with supporting tank, anti-tank, anti-aircraft, reconnaissance and rocket battalions. In time of war its theatre of operations would be Denmark. In addition to this, the navy itself has a small number of swimmer/diver teams.

CZECHOSLOVAKIA

After the Soviet invasion in 1968, the Czech People's Army, Ceskoslovenska lidova armada, was no longer regarded as trustworthy and was drastically reduced in strength. The 22nd Airborne Brigade was reduced to the size of a single regiment with four battalions, one active, one reserve, one training and one dedicated to special operations. The regiment is based at Prosnice.

HUNGARY

Hungary's sole elite force is a single airborne battalion of approximately 400 men.

ROMANIA

Romania, which has now banned the Communist Party entirely, has for many years only been a nominal member of the Warsaw Pact and rarely participates in joint manoeuvres. Its army includes a single airborne regiment, the 161st, based at Buzau, two alpine brigades, the 2nd at Brasov and the 4th at Cuerta de Arges, and a naval infantry battalion at Ciurga. The latter only has a defensive role, however. The real elite are the two alpine brigades for Romania has a long tradition in mountain warfare and these are undoubtedly amongst the most skilled in the world.

BULGARIA

Bulgaria has three companies of naval infantry which regularly exercise in amphibious landings alongside their Soviet counterparts in the Black Sea Fleet. In addition there is a single airborne regiment, a mountain brigade of unknown strength or composition, and an equally unknown number of small Spetsnaz-type units which are known in the past to have been employed by the Russian KGB for assassinations in the West.

Additionally, all Warsaw Pact nations have strong internal security or 'secret police' forces, although as in Romania which has already disbanded its Securitate, these may not last much longer.

Other Nations

Peruvian troops demonstrate to a fascinated audience the technique of abseiling to the ground from helicopters.

ARGENTINA AND OTHER SOUTH AMERICAN COUNTRIES

Since the Falklands War and the overthrow of the military junta, the Buzo Tactico appears to have been replaced by a purely counter-terrorist group called Halcon (Falcon) 8. Bolivia has a small special forces' training centre. Brazil has a full parachute brigade equipped and trained American-style plus the counter-terrorist Talon group. Chile has a single paratroop battalion and six companies of special forces including police and mixed army/police counter-terrorist units trained by American instructors; Peru has essentially the same structure. Colombia, facing special problems in the drug war, has a police special operations group and a similar army unit which is specifically dedicated to stamping out kidnapping and extortion. The Dominican Republic, Ecuador, Honduras and Venezuela each also have small army special forces groups. The status in Paraguay is unknown.

US Green Beret advisors helping special forces in Ecuador over a jungle assault course.

AUSTRALIA, NEW ZEALAND AND OTHER FAR EASTERN AND OCEANIC COUNTRIES

Modelled on the British pattern, a single Australian Special Air Service company was formed in 1957. Serving for a while as part of the Royal Australian Regiment from 1960, in 1964 it was granted independence and increased to the stature of a full regiment with two 'Sabre' squadrons, a headquarters/base squadron and a signals squadron, which saw active service in Borneo. From 1962 to 1971 the Australian SAS was involved first in a training and then in an infiltration, reconnaissance and ambush role in Vietnam, a third squadron being added in 1966. The 'Oz' SAS were influential in the development of the US 'Recons' and frequently fought alongside the Green Berets and SEALs in this most difficult war. The second 'Sabre' squadron was disbanded when Australia pulled out of Vietnam but was reinstated in 1982.

The regiment's members are recruited from volunteers from other army units and have to undergo exactly the same training as their British counterparts although there is an increased emphasis on jungle warfare and amphibious techniques. Today, the Australian SAS is principally a counter-terrorist formation with responsibility for safeguarding the nation's offshore oil rigs, alongside the 1st Commando Regiment which was raised in 1980 with amphibious operations particularly in mind.

The New Zealand SAS was first raised in 1954 to join the British and Rhodesians in Malaya and shared the Australian experience in Borneo and Vietnam. Known since 1963 as the 1st Ranger Squadron, NZSAS, it currently consists of a headquarters and five active 'Sabre' companies whose duties are principally counter-terrorist.

China has three known airborne divisions based loosely on the Soviet model in its People's Liberation Army, but there may be more. Each totals approximately 799 officers and 8,300

ABOVE Members of the Australian SAS enjoy a break during the fighting in Borneo.

OPPOSITE A New Zealand paratrooper opens his 'chute at 2,000 ft (600 m).

men organized in three parachute regiments plus headquarters, artillery and signals battalions and anti-aircraft, reconnaissance, 'guard', engineer and chemical warfare companies – the Chinese do not bother with a euphemism for the latter. There is a guard company in all Chinese army divisions, entrusted with security. One obvious deficiency in the Chinese airborne formations is the lack of a specific anti-tank component.

Taiwan, the island formerly known as Formosa to which General Chiang Kai-Shek's forces retreated after their defeat by the communists in 1949, has been propped up with decreasing determination by America and its armed forces are almost exclusively US-trained and equipped. Since American interest in propping up the regime has dwindled close links have been forged with both Israel and South Africa, equal specialists in counter-insurgency operations. The current disproportionately large Taiwanese army of more than 250,000 men includes two parachute brigades, a number of SEAL-style combat frogmen trained in free-fall parachuting and a Long Range Amphibious Reconnaissance Commando similar to the USMC's Recons.

Both North and South Korea have maintained armies equally disproportionate in size since the war of 1950–53 and both maintain strong special forces units. The North Korean elite forces which come under the control of VIII Special Operations Corps are the strongest, an incredible 80,000 men in 22 brigades spread between the army's eight Corps. Principally trained Ranger-style as light infantry after the statutory month-long 'boot camp', the highly aggressive North Koreans also learn amphibious and mountain warfare techniques and many of them also master parachuting and scuba diving. South Korean security forces regularly intercept reconnaissance and sabotage teams coming off the beaches and through the demilitarized zone, and routinely shoot to kill – a far cry from the 'polite' deterrent tactics used

against Spetsnaz by the European NATO nations! National service in North Korea lasts for seven to 10 years, resulting in a very high level of proficiency compared with the conscript armies of other countries. South Korea has seven special forces groups moulded on the lines of the Green Berets and similarly trained and equippped. In addition to normal qualifications, each member has to be a black belt in the Korean equivalent of karate. The country also has a police counter-terrorist unit backed by the army's 707th Special Missions Battalion.

India formed its first parachute brigade in 1941 and by 1944 had a complete division incorporating British and Gurkha regiments. After independence and partition with Pakistan in 1947 this was renamed the 50th Indian Parachute Brigade which survives to this day. It saw action against Pakistan in the 1965 war and currently consists of the 9th and 10th Parachute Commando Battalions, each of three rifle companies plus supporting artillery, engineer and signals detachments. All its members are volunteers and after basic training have to complete a course culminating in five jumps. Pakistan has a slightly larger parachute-trained brigade of three battalions, each four companies strong plus supporting services.

Thailand's famous 'ninjas', with their close-fitting black coveralls and hoods, expertise in Thai-style boxing with hands and feet, Ranger and airborne training and experience in jungle survival and small unit tactics, are amongst the most formidable in the world. They need to be with hostile Burma and Kampuchea to either side of them. The first Thai special forces battalion was formed in 1963, but during 1982–83 they were expanded to two brigade-sized 'divisions', the 1st and 2nd Special Forces Division, each consisting of three regiments of about 1,000 officers and men. All their members are highly motivated volunteers, some of whom train alongside the US SEALs. An oddity about their training programme is that, alongside Britain,

**South Korean paras
hurry to leave their
aircraft so they will land
in a tight formation on
the ground.**

Thailand retains balloons for the first static-line jump.

Other Far Eastern elite units include two Indonesian airborne brigades subordinated to Special Warfare Command, plus army, navy and police counter-insurgency units; Japan's 1st Airborne Brigade which, although parachute-trained, is heliborne; the Philippines' army, air force and police light reaction and special operations groups; and Singapore's counter-terrorist Police Tactical Team.

AUSTRIA AND OTHER EUROPEAN AND SCANDINAVIAN COUNTRIES

The small army of neutral Austria includes a number of alpine units trained in Ranger and mountain and arctic warfare techniques plus the counter-terrorist Gendarmerieensatz-kommando better known as 'Cobra' force which is the equivalent of Delta or GSG9.

Like Austria, neutral Sweden knows it occupies a special place in the minds of Warsaw Pact planners because an invasion of Norway would not be feasible without violating that neutrality any more than an invasion of West Germany would be without marching through Austria. Both countries are caught in geopolitical traps if ever the worst should come to the worst, and the attentions of the Soviet submarines and Spetsnaz teams off the Swedish coast have been much publicized. Sweden takes its neutrality seriously and has some of the most modern guns, tanks and aircraft in the world, almost all 'home grown' to protect the country's non-alignment policy.

National service is compulsory for up to 15 months and selected volunteers enter one of the Swedish special forces companies, of which there are three types. (All officers are career soldiers, either active or in the reserves.)

As you would expect, Austrian troops are superb skiers, whether operating in the wilds (opposite) or hitching a tow behind a Volvo oversnow vehicle (below).

First is the Fallskärmjagere or Parachute Ranger Company which would be deployed in small reconnaissance patrols of four to six men, all skilled skiiers and mountaineers as well as free-fall parachutists. Second is the Kustjärgarskolan or Coastal Ranger Company trained in every aspect of coastal reconnaissance and patrol, small boat handling, etc. Finally there are the navy's swimmer/diver teams who are trained in the assault role as well as the usual tasks of beach reconnaissance, surveillance and under-water demolition.

Finland, outwardly neutral but in fact implac-ably hostile to the Soviet Union ever since the Winter War of 1939–40, does not openly admit to possessing special forces but the calibre of its troops and their experience of arctic warfare means that they are all men to be reckoned with. It is similar with the Swiss 'citizen army' which knows that if war comes it is every able-bodied man's duty to defend its centuries-old policy of neutrality, but the Swiss army is known

OPPOSITE Swedish troops with a 90 mm recoilless anti-tank gun on an oversnow vehicle. **ABOVE** Lurid helmet devices are used during live firing exercises by the Turkish army to prevent casualties; the weapon is a 7.62 mm FN FAL rifle.

to contain a mountain-trained Parachute-Grenadier Company. Among the usual subjects taught to the special forces of other nations, the Swiss also study avalanche formation, not just so they can help the civil defence authorities in the event of a disaster but so they could delib-erately trigger avalanches on the heads of an invader.

The Turkish special forces are also unusual. Coming under the command of the army's Special Warfare Department, they are specially trained to organize and lead guerrilla forces within Turkey itself in the event of an invasion. Any attempt to pin down their organization is, for obvious reasons, impossible. Much the same applies to Yugoslavia, whose forces have bitter experience of the long and deadly guerrilla war they fought against the Germans from 1941 to 1945. Fully trained in all aspects of mountain and arctic warfare, they could also be infiltrated in small ambush and sabotage teams behind enemy lines.

THE MIDDLE EAST

When one thinks of elite units in the Middle East, the first country which springs to mind is Israel, followed immediately by the word 'Entebbe', which was the classic special forces' hostage rescue operation of all time. In the early hours of the morning of 27 June 1976, sleepy-eyed passengers boarded Air France Airbus Flight 139 at Athens, its second stopping point en route from Bahrein to Paris via Tel Aviv. No-one took any notice of the German lawyer and his girlfriend in the first-class section or a pair of nondescript Arabs in tourist class until the woman drew a grenade shortly after take-off and told everyone to stay seated while her companion produced an automatic pistol and commandeered the flight deck. The two Arabs started positioning explosive charges disguised as boxes of chocolate around the cabin. The two first-class passengers were actually members of the German Baader-Meinhof gang, the two Arabs of the PLO, and their demand was the release of 53 fellow terrorists imprisoned in France, Germany and Israel.

The aircraft refuelled at Benghazi in Libya then flew on to its pre-arranged destination, Entebbe in the African state of Uganda, then ruled by the manic self-appointed Field Marshal Idi Amin. Amin presented himself to the world and to the terrified hostages as a mediator, but in fact Ugandan soldiers mingled amicably with PLO guerrillas at the airport, where the 12 crew members and 246 passengers were herded into the old terminal building. Over the next few days most of the passengers were released, but apart from the Air France crew who refused to leave, the remaining 104 hostages were those with Jewish names or Israeli passports.

The Israeli Defence Forces had not been inactive and while the Knesset, or cabinet, debated what should be done, various contingency

C-130 Hercules transport aircraft which carried Israeli paras into action at Entebbe and in many more conflicts.

plans had been drawn up. Israeli secret service men slipped into Uganda from neighbouring Kenya and adopted the identities of airport workers to spy out the strength and dispositions of the terrorists, while a mixed force of some 200 men from the Israeli paras and elite Golani Brigade assembled at Ophira air force base near Sharm el-Sheik to rehearse the operation in and around a hastily constructed replica of the Entebbe terminal. They were accompanied by Ulrich Wegener, commander of GSG9, as a very interested observer.

Approval for the rescue operation, code-named 'Thunderbolt', finally came through on 3 July, by which time many of the hostages were in a pititful state through lack of decent food or potable water. They embarked in four Hercules transports which flew at low level through a thunderstorm over the Red Sea and then across Ethiopia to avoid radar detection, while above them a Boeing 707 converted into an aerial command post with sophisticated communica-

ABOVE AND RIGHT
Scenes from the film
Operation Thunderbolt
which recreated with
considerable accuracy
the Israeli hostage
rescue mission at
Entebbe.

9mm UZI

Named after its designer, Lieutenant Uziel Galil, the 9mm UZI sub-machine gun was developed after the War of Independence to replace the motley collection of non-standardized smallarms which the IDF had acquired from a variety of European sources. It has since become one of the most widely used SMGs in the world, being favoured by many parachute and special forces groups and even carried by the United States President's Secret Service bodyguards. In Belgium it is made under licence by Fabrique Nationale.

The basic weapon, which fires a 9mm Parabellum cartridge, is largely of inexpensive stamped steel pressings, spot-welded together, although the barrel, breech and magazine receiver are machined components. Overall length has been kept down by having the 25- or 32-round box magazine inserted through the pistol grip, which also makes reloading faster, especially in the dark. Another clever feature of the design is the internal sand and dust grooves which collect any dirt entering the weapon and prevent it clogging the firing mechanism. The weapon can be fitted with either a fixed wooden stock or a folding tubular stock.

Weighing only 9lb (4 kg) fully-loaded, the UZI has a rate of fire of 600 rounds per minute and an effective range of about 100 yards (90 m). Israeli Military Industries have also developed a mini-UZI with a folding stock and shorter barrel (7.75 in or 197mm instead of 10.25 in [260mm]) for covert operations, since it can be concealed almost as easily as a pistol. Range, of course, suffers, but this is not usually important in a hostage situation. The mini-UZI showed its value when Israeli commandos stormed Beaufort Castle in Lebanon in 1982, a PLO stronghold. The walls had to be scaled using ropes and grappling hooks, and this small weapon could easily be slung out of the way but instantly retrieved as each man reached the parapet.

UZI with folding metal stock.

tions equipment flew so close to a scheduled commercial airliner heading to Nairobi that their two radar blips merged as one. A second 707 fitted out as an emergency hospital also flew to Nairobi.

On the ground at Entebbe the first three Hercules touched down unmolested because telephone lines from the control tower had been cut and the operators could not find anyone to give them instructions. They taxied to the end of the runway and the assault troops began fanning out to protect the perimeter from counter-attack by the nearby Ugandan troop garrison and to immobilize Mig-17 fighter aircraft parked nearby which could have shot down the Israeli aircraft on their return.

The fourth Hercules landed and taxied round in front of the old terminal building. Its rear ramp came down and out rolled an exact copy of Idi Amin's black Mercedes. He often conducted snap inspections and talked to the hostages, so the guards were not unusually alerted. Ugandan troops at the time were poorly disciplined so the sound of gunshots from the new terminal were virtually ignored. However, sentries in front of the old terminal quickly spotted the fact that the Mercedes was packed with gunmen and opened fire. They died quickly and the Israeli Commandos stormed into the building, screaming in Hebrew at everyone to lie flat. The two German terrorists were among the first fatalities, but the leader of the Israeli assault force, Lieutenant-Colonel Yonatan Netanyahu – popularly known just as 'Yoni' – was also killed, along with three other soldiers and three of the hostages who stood up instead of lying down, while others were wounded. The latter were flown to the medical 707 at Nairobi while the remaining three aircraft made their triumphant flight home.

The Israeli Defence Forces (IDF) were forged in the fires of the War of Independence in 1948 following the British withdrawal from Palestine and have been constantly at war ever since, making them the most combat-experienced troops in the world. From the invasion of Egypt alongside the British and French in 1956 through the War of Attrition and the Yom Kippur War of 1973 to the continuing confrontation in Lebanon, Israeli troops have honed their skills while the country's defence industries have produced a wide range of indigenous weapons to safeguard against the threat of embargo from other nations. Apart from the major conflicts, Israeli airborne and commando forces have conducted many interdiction strikes at targets in neighbouring Arab countries, as in 1968 when heliborne troops seized Beirut airport and destroyed all Arab airliners present in reprisal for PLO attacks on El Al aircraft in Europe.

Israeli para with Galil 5.56 mm ARM assault rifle.

The first ad hoc Israeli parachute unit was formed during the War of Independence and by 1955 had been expanded to the size of a brigade, the 202nd. It was a battalion from this unit which in 1956 parachuted in to seize the Mitla Pass in advance of the ground forces heading towards the Suez Canal. Subsequently the Israeli parachute force has been expanded to six brigades, three active and three reserve.

There is keen competition within Israel's citizen army to be selected for the paras, despite the gruelling six-month training programme. All volunteers first undergo two months' basic, much of the training taking place in the field to toughen the recruits and teach them endurance during long forced marches across rough terrain. Weapons training is also intensive, the paras' favourite weapon being the UZI submachine gun. This phase is followed by three months' specialist training, during which each soldier learns a particular skill such as machine gunner, signaller, medic or driver and also practises assaults using armoured personnel carriers and helicopters. Finally comes jump training, culminating in five static-line drops, after which the successful recruits are awarded their wings and join an active brigade. Even here training never ceases, and many paras go on to learn free-fall parachuting techniques and other skills to suit them for entry to one of Israel's special forces groups.

The IDF has a number of special forces designated Sayeret (reconnaissance). Sayeret Almond is part of Southern Command and specializes in Long Range Desert Group-style patrols with jeeps deep into the desert. Sayeret Carob is part of Central Command and specializes in mountain warfare in the Sumerian and Judean hills. Sayeret Egoz, part of Northern Command, is also mountain-trained, as is Sayeret Golani, the reconnaissance troop of the elite Golani Brigade. Sayeret Hadruzim is unusual in that its personnel are all Druze Moslems; it also operates as part of Northern Command. The only thing known about Sayeret Matkal is that it is entrusted with the most sensitive operations behind enemy lines. Sayeret Orev is the special reconnaissance unit of the para brigades and is unusually given a dedicated anti-tank role as well. Sayeret Shaldag is an infiltration and sabotage unit; Sayeret Shimson is a naval infantry force trained in reconnaissance from the sea; and finally Sayeret Tzanhim is a shock troop specially trained in rapid assault tactics. The IDF also has a number of Naval Commando units whose men practise scuba diving and free-fall parachuting as well as conventional amphibious warfare techniques. Their training is even more intense than for entry into the paras and lasts 12 months. Last but not least, since Entebbe a special permanent counter-terrorist force trained in all aspects of hostage rescue has been formed, known just as Unit 269.

Several Arab countries also have paratroop and special forces units, including Egypt, Jordan and Syria. The first Egyptian parachute unit was formed in 1959 and first saw action in Yemen in 1965. Since then they have been expanded to two Parachute Brigades and two heliborne Air Assault Brigades, each about 3,000-strong, plus a Marine Assault Brigade and seven 1,000-strong independent Commandos. Originally the parachute forces were trained by Soviet instructors but in recent years their teachers have been American, as are their parachutes. In addition to these regular forces Egypt also has its own small 250-strong counter-terrorist force, Sa'Aga ('Thunderbolt'). It has seen action three times: at Luxor in 1975 when it succeeded in rescuing all the passengers from a hijacked Boeing 737; unsuccessfully at Nicosia, Cyprus, in 1978 when it got into a shoot-out with Cypriot security forces while attempting to rescue hostages from a hijacked aircraft and lost 16 dead; and at Luqa, Malta, in 1985 when 59 out of 98 people aboard another 737 unfortunately died. Sa'Aga, which is trained by GSG9, has not, therefore, been terribly lucky.

Jordan did not form its first parachute unit until 1963 and currently has a brigade of three 500-man battalions, each subdivided into three companies. Among the toughest troops in the Middle East, they are all Bedouins owing personal loyalty to the Royal Family. They are specially trained in guerrilla and counter-insurgency warfare and scored a notable success in 1976 during a shoot-out with PLO gunmen holding guests hostage in the Intercontinental Hotel.

Syria's first parachute battalion was raised in 1958 and there are now three airborne commando battalions, each of three companies. They are trained in helicopter assault as well as parachuting and were successful in capturing the Israeli observation post on Mount Harmon at the beginning of the Yom Kippur War, although they were later thrown out by a determined counter-attack. Since 1982 they have seen almost continuous action against the Israelis in Lebanon.

SOUTH AFRICA

Existing in a permanent state of siege and having fought wars in Angola, Rhodesia and Namibia over the last couple of decades, the South African army has developed into the most formidable counter-insurgency force in the world. It has a single Parachute Brigade based at Bloemfontein, a heliborne Para-Commando at Kroonstat, but the truly elite unit is the Reconnaissance Commando, also based at Bloemfontein. Volunteers for the 'Reccondo' come from all branches of the armed forces, not just the army, and have to go through a gruelling 42-week training course which fewer than ten per cent manage to pass. There are two intakes of about 700 volunteers a year, the courses overlapping, and at the end of this 12 months only about 100 men will be accepted into the unit.

Training is conducted in the wilds of northern Zululand and includes arduous forced marches wearing full kit by night and day across the veldt, swimming and rock climbing. All the men have to learn free-fall parachuting and particular emphasis is placed on tracking skills and laying ambushes. A high level of marksmanship is demanded as well as skill in unarmed combat and the use of the 'Warlock', the South African commando knife. Many members of the Reccondo are former soldiers of the Rhodesian SAS and Selous Scouts, which were both disbanded when the country became Zimbabwe, and a few are disaffected Angolan mercenaries. The Reccondos specialize in hot pursuit missions, tracking guerrilla groups across the borders into neighbouring countries on occasion. They are probably the finest light infantry on the globe.

ABOVE South African paras during a static line jump. **RIGHT** A member of the Reconnaissance Commandos, probably the finest light infantry in the world.

Index

1 Italicized page numbers refer to illustrations.

2 Numerical entries are entered as if spelled out, eg 82nd Airborne is entered as if it were written as Eighty-Second.